ADHESIVES AND GLUES

how to choose and use them

ADHESIVES AND GLUES

how to choose and use them

Robert S. Miller

Franklin
CHEMICAL INDUSTRIES

2020 Bruck Street   P.O. Box 07802   Columbus, Ohio 43207

Manufactured in the United States of America

Text prepared and book designed by Robert Scharff & Associates, Ltd.

Current Printing (last digit) 10   9   8   7   6   5   4   3   2   1

S & S Publishing Co., Box 192, New Ringgold, Pa. 17960

Acknowledgments

The author is deeply indebted to many individuals who helped in many ways to create this book.

A special thanks goes to Dr. Robert F. Snider, whose expertise, previous writings, and assistance in the wood gluing sections were invaluable. Jack Gerard and Larry Clark also contributed much to the chapters on wood and furniture gluing and repair.

Without a doubt, the Elastomers Laboratory section of Franklin Chemical Industries under the direction of George Butts and Denny Doyle have, through the years, provided the writer with much guidance and insight into both the technical and practical use and application of adhesives and glues.

There are many others both inside Franklin Chemical Industries as well as outside the company to whom we owe a great deal of gratitude for their direct and indirect contributions in bringing this book to fruition. Our thanks goes out to all for their help, guidance, and patience.

**Library of Congress Cataloging
in Publication Data**

Miller, Robert S.
    Adhesives and Glues
    how to choose and use
    Includes index.
                                80-51745

Contents

Preface

Being a leader in the manufacturing of adhesives and glues, Franklin Chemical Industries saw a need to more thoroughly inform the consumer about the various types and uses of adhesives in the home and workshop. Although there have been books written on certain phases of gluing, never before has such a complete book as this been published.

Our book covers a wide range of subjects to prepare the amateur do-it-yourselfer for choosing the proper adhesive or glue for the job at hand. It deals with selecting from among the many types of available wood adhesives, using wood adhesive in joint design, and conditioning and preparing wood for gluing. The book goes even further into detail on the subject of using glues in woodworking in the chapter on gluing and clamping. Included in this chapter are sections on clamping devices, wood gluing procedures, glue assembling of furniture, decorative lamination, and veneering. Another chapter takes the use of adhesives and glues on furniture one step further by describing the many repair techniques used. This involves such items as employing caster inserts to mend split legs, tightening joints with thread or cloth, plugging holes with dowels, tightening joints with wedges, and much more.

The book also includes the contact adhesives: the types, how to laminate using contact adhesives, and ideas for using plastic laminates in many practical and imaginative applications. Subjects also covered are the installation of walls and ceilings with adhesive—types of construction adhesives, wall paneling materials, as well as installing ceramic tile walls, brick veneer and simulated brick walls, and ceiling tile. In addition, there is an entire chapter dealing with the methods for putting down floors with adhesives—the types of adhesives used for bonding flooring materials, subfloor systems and their preparation for bonding, resilient flooring, hard floor materials, and carpeting. Yet another chapter describes the many kinds of specialty adhesives, such as cyanoacrylates, epoxies, and so on, and their use in specific bonding jobs. The final chapter will aid the layman in determining the reasons why an attempted bonding job might not have succeeded and suggests ways to overcome problem areas when performing an adhesive operation.

I have known the author of this book, Robert S. Miller, for 15 of his 28 years in the adhesives industry. He has worked in both the industrial and consumer fields of adhesive and glue applications. Having been with Franklin Chemical Industries for the past 12 years, Bob has served as marketing manager of our Consumer Products Division and is presently general manager of our Contract Packaging Division. He holds a B.S. degree in marketing from the Ohio State University and has done graduate work at Syracuse University. Bob is currently participating in a graduate program leading to an M.B.A. at Ohio University. I feel that the author's wide experience in the construction as well as the adhesives industry qualifies him from a practical standpoint to write in a layman's language about construction and craft applications of adhesives and glues.

It is our sincere hope that this book will enable the reader to undertake any sort of gluing operation—and do it with confidence and knowledge.

L. Thomas Williams, Jr., President
Franklin Chemical Industries

Introduction to Adhesives and Glues 1

Glues and adhesives have been around for a long time. The ancient Egyptians used them to fasten fine veneers and ivory inlays to wood bases. The early Greeks and Romans employed cements and mastics to apply their beautiful ceramic tiles to floors and walls. Many of the famous cabinetmakers of yesteryear—Chippendale, Sheraton, the Adam Brothers, Hepplewhite, Duncan Phyfe—all used glue to hold their furniture pieces together.

The glue that the earlier furnituremakers used was made from the hides, horns, and hooves of animals, boiled down to a jelly and then allowed to harden. This dried mass was then broken into flakes or ground into coarse powder. When used, it was mixed with water and gently heated in an iron glue pot. This glue was hard, brittle, and brown in color. It was not waterproof and often left a stain on wood. But, in spite of its faults, this glue was all that was available until World War I, when casein glues—made of milk—and nitrocellulose glues—the same material that is used in the making of gun powder—became available.

During the 1930's, the plastic resin glues—urea-resin and resorcinol resin—made their appearance on the market. The latter was the first really waterproof adhesive, and it played a very important part in the building of World War II's famed torpedo boats. While many new adhesives—epoxies, acrylonitriles, neoprenes—were developed during the war, none reached the market place until the late 1940's or early 50's. Since that time, progress in adhesives technology has been phenomenal, and it is still advancing at an accelerated rate.

Match a modern adhesive to the job and you can bond almost any material. While some adhesive products are used for specific bonding problems, there are also multipurpose types which can be used for many and varied home and workshop jobs. Not only do adhesives differ in the amount and kinds of purposes which they serve, but they also differ in the conditions under which they can be used. For instance, temperature is an extremely important factor to consider. While there are those adhesives which can be used at practically any temperature, there are also adhesives which must be used within specific temperature ranges. Some adhesives will not be able to bond if the surfaces to which they will be applied are not properly prepared and cleaned; others need no special preparation. Some adhesives must be mixed before they are used, while others come ready-mixed. Finally, when working with some adhesives it will be necessary to clamp your workpieces; on the other hand, some will not need the additional pressure supplied by clamping.

In any case, to be certain of the most efficient bond: (1) Use the proper adhesive or glue for your specific job; and (2) follow the manufacturer's directions for application. Always remember that these recommendations given by the adhesive producer will provide the most successful bonding results and it would be wise to follow them. However, it will be necessary to use your own judgement when deciding upon the right type of adhesive or glue for a specific job. Although the manufacturer's suggestions for the uses of a particular adhesive are usually

1

valid, there may be instances when a product designed for a specific application will be better for the job at hand than a multipurpose adhesive which is also recommended for the job. That is, most glues today are designed for use on a particular material or class of absorbent or nonabsorbent materials. Thus, polyvinyl acetate (white) glue is excellent for woodwork, but it does have other absorbent materials (paper, leather, and fabric) for which it can be used. However, white glue cannot be used to bond glass to metal or steel to plastic, etc. Nonabsorbent materials like ceramics and metals can be bonded with cyanoacrylate, but woods and other absorbent materials cannot.

A final consideration to keep in mind is the cost of a particular bonding product. Although an expensive adhesive or glue may serve your purpose, there is also the possibility that a less expensive product may work just as well.

The wisest way to choose an adhesive or glue is to find one which is designed particularly for the job you need to do. For instance, if you were to buy a glue specifically designed for wood and similar applications, you may find that the container states that it can also be used for ceramics. The glue probably would work on ceramics, but it might not work as well as one designed for ceramic use only. In addition, the claims on a container of adhesive can often be interpreted in several ways. For example, there is no real dividing line between a product that is truly waterproof, stainproof, or heat-resistant and one that is not. What amount of immersion in water will destroy a glue bond? At what temperature will an adhesive fail? It is probable that the manufacturer knows these limitations, but the average do-it-yourselfer does not. It is a good idea to use widely known adhesive products; therefore, you can feel confident that the manufacturer's claims will hold true.

There really are not as many adhesives on the store shelves as it seems to the occasional buyer. Often the same glue appears under different names. For example, aliphatic glue is called "Titebond" by Franklin Glue Company and "Carpenter's Wood Glue" by Borden (Elmer's). Cyanoacrylate appears under such

Fig. 1-1: Both trade names and type names identify various adhesives.

Fig. 1-2: The adhesive's label and container lip often carry important information.

names as Eastman 910, Flash, Duro's, Superglue 3, Quickfix, Speedweld, Zip Grip, and others. With some adhesives (Fig. 1-1), however, trade names plainly identify a particular adhesive. The words "epoxy" and "contact" are good examples. Similarly, adhesives used to fasten paneling to walls or studs usually have the words "panel" or "construction" in their names or descriptions.

The adhesive label on the container also carries other important information (Fig. 1-2). For instance, some solvent-based adhesives may be extremely flammable and contain dangerous toxic fumes. As a result, the Consumer Product Safety Commission placed a ban on consumer sales of extremely flammable contact cements. You can recognize these glues on your shelf by required warnings, such as "Danger, Extremely Flammable, Keep Away From Children." If you have some of these still on your shelf, do not smoke, turn off all pilot lights, and work in a well-ventilated room when using them.

Many manufacturers now produce a water-based version of flammable adhesives. This is especially true of contact cements. The flammable types dry faster than the water or emulsion systems because the solvents evaporate faster. However, there are now nonflammable solvents, utilizing chlorinated solvent systems, that will dry equally as fast as their flammable counterparts. These are becoming quite common in contact cements and flooring adhesives. All three types are nearly equal in effectiveness as adhesives. The nonflammable types, though, eliminate both the fire hazard and the other hazards associated with volatile solvent systems. Caution should be exercised in the use of the non-flammable chlorinated solvent systems, and prolonged inhalation should be avoided. Tools used with the water-based type can generally be cleaned with soap and water while the glue is still wet.

When selecting an adhesive for a project, there are two major factors which, when taken into consideration, will greatly simplify the task. They are:

1. Type of material to be fastened. Are the materials to be joined porous or nonporous? Wood, cork, paper, leather, cardboard, and cloth are considered porous materials, while metal, glass, ceramics, porcelain, tile, and most plastics

are nonporous. Are the parts easily clamped together? Are you trying to join two dissimilar materials such as wood and plastic? Are the materials rigid or flexible? Are the materials being bonded smooth or rough?

2. Physical conditions and characteristics. Will the completed bond be subjected to heat, cold, or moisture? Are there large voids that require a gap filling function by the adhesive? To what kind of stress will the finished joint be subjected? Is the glue flammable? Are the materials structurally sound or are there internal weaknesses that will affect the strengths? What is the color of the adhesive when it is dried? Will any surface preparations be required to provide an adequate bonding surface? Is drying time important? Who is the end user of the bonded item and where is the bonding taking place? If you are repairing a child's toy or working in an area with poor ventilation, you do not want to use an adhesive with a toxic effect.

With the answers to these important questions in mind, you should be able to go a long way toward selecting the proper adhesive or glue for your putting-it-together task. But, taking a look at the various modern day adhesives and glues and how to use them, let us attempt to clear up a point of confusion in the minds of some users: What is the difference between the terms "adhesive" and "glue"?

In today's world (and in this book), the two terms are used almost interchangeably. At one time, the word "glue" referred only to products which were derived from an organic material and were used only on porous materials. The word "adhesive" was used only to describe products derived from a synthetic resin and used on nonporous surfaces. Nowadays, however, the majority of glues are made from synthetic resin bases, and most can be employed for other jobs in addition to their primary function of gluing wood and similar porous materials. Incidentally, the use of all adhesives is generally referred to as "gluing."

Wood Glues and Adhesives 2

Gluing is done extensively in woodworking and in the production of various types of wood products. Modern wood adhesives and techniques for using them vary as widely as the products made, and developments have been substantial in recent years. In general, however, it still remains true that the quality of a glued joint depends upon the kind of wood and its preparation for use, the kind and quality of the adhesive, the details of its gluing process, the types of joints, and the conditioning of the joints. Depending on the adhesive used, application conditions also affect the performance of the joint to a greater or lesser extent.

WOOD ADHESIVES

Modern wood adhesives, as a group, are easy to use and, if properly applied, will make a joint stronger than the wood itself. Wood glues are generally available in three forms: (1) Ready-to-use (no mixing required); (2) water-mixed (powder adhesive must be mixed with water); and (3) two part (the two parts must be mixed together).

Ready-to-Use Adhesives

Ready-to-use glues—the most popular wood adhesive with both commercial and do-it-yourself consumers—are widely used for permanent joint, edge, face, assembly, and laminating gluing. The most common of these glues are the liquid hide, aliphatic, and polyvinyl acetate glues. All of these glues set by loss of water from the glue line.

Essentially the gluing operation consists of applying a liquid adhesive (a suitable polymer) dispersed in enough water to give good spreadability and then pressing the parts tightly together until the glue sets. During this setting process, the glue and water penetrate the pores of a thin wood surface layer. As the water goes through the pores, the glue is retained on the wood fiber walls. The adhesive gains strength as the water leaves the adhesive film. Speed of set, assembly time, and depth of glue penetration are dependent on the speed of water removal. Strength, amount of wood failure, resistance to water, humidity, solvents, and heat are dependent on the nature of the polymer.

Many properties of an adhesive contribute to its successful practical application. The three properties of ready-to-use adhesives that are important to the home craftsman are:

1. **Viscosity.** Viscosity is a measure of the resistance to stirring or thickness of a liquid. It is measured in centipoises or poises (a poise equals 100 centipoises). The higher the viscosity (the greater the centipoises count) reading, the thicker the liquid. Actually, with most adhesives, viscosity gives an indication of the ease of application and coverage obtainable with a brush spread. For instance, a polyvinyl acetate adhesive with a viscosity of about 3,000 cps. will brush much more easily than a heavier liquid hide glue with a viscosity of up to 5,500 cps.

2. **Wet Tack.** Some glues, such as the aliphatic series, have a built-in tack. Although this wet tack retards brushability, it does not affect the spreadability by

5

other means. Tack prevents a tight dowel from easily scraping off the glue from the inside of a dowel hole. It also holds light parts, such as corner blocks, in place.

3. **Temperature.** Temperature affects glues in different ways. For example, lowering the temperature of a liquid hide glue toward its gel point (usually near 70°F) will cause it to thicken rapidly. Aliphatic and polyvinyl resin glues are less affected (though they do thicken) as the temperature is lowered toward freezing. Also, raising the temperature above room temperature affects glues differently. Liquid hide glues are thinned by raising their temperature, although prolonged exposure to elevated temperatures, such as 140°F, will decrease their strength. Aliphatic glues and some polyvinyl acetate glues become thinner as the temperature is raised until they reach 120°F; at higher temperatures they become irreversibly thick.

Most ready-to-use glues (Fig. 2-1) are available in 2, 4, 8, 16, and 32 ounce sizes; some can be had in a 128 ounce size.

Now, let us take a closer look at the major group of ready-to-use glues.

COMPARISON OF TYPICAL READY-USE ADHESIVES

	Aliphatic Resin Glue	Polyvinyl Acetate Glue	Liquid Hide Glue
Appearance	Cream	Clear white	Clear amber
Viscosity (poises at 83°F)	30-35	30-35	45-55
pH*	4.5-5.0	4.5-5.0	7.0
Speed of Set	Very fast	Fast	Slow
Strength (ASTM# Test)	All three easily exceed government specifications of 2800 pounds per square inch on hard maple. On basis of percent wood failure, aliphatic best, liquid hide next, polyvinyl acetate third.		
Stress Resistance†	Good	Fair	Good
Moisture Resistance	Fair	Fair	Poor
Heat Resistance	Good	Poor	Excellent
Solvent Resistance‡	Good	Poor	Good
Gap Filling	Fair	Fair	Fair
Wet Tack	High	None	High
Working Temperature	45°-110°F	60°-90°F	70°-90°F
Film Clarity	Translucent but not clear	Very clear	Clear but amber
Film Flexibility	Moderate	Flexible	Brittle
Sandability	Good	Fair (will soften)	Excellent
Storage (shelf life)	Excellent	Excellent	Good

*pH—glues with a pH of less than 6 are considered acidic and thus could stain acid woods such as cedar, walnut, oak, cherry, and mahogany.
#ASTM—American Society of Testing Materials.
†Stress resistance (cold flow)—refers to the tendency of a product to give way under constant pressure.
‡Solvent resistance—ability of finishing materials such as varnishes, lacquers, and stains to take over a glued joint.

Liquid Hide Glues. The liquid hide or animal glues are actually a ready-mixed version of one of the oldest types of wood glue (if not the oldest) which furniture and cabinetmakers once cooked or boiled in a pot. Still made from animal hides, bones, and tendons, liquid hide glue is excellent for interior furniture construction and repair. It makes a joint stronger than the wood itself. The shear strength on hard maple will exceed 3,000 pounds per square inch (psi). Liquid hide glues are suitable for gluing together many different types of materials in addition to wood, such as hardboard, chipboard, leather, cloth backed imitation leather,

Fig. 2-1: Ready-to-use glues are usually available in 32, 16, 8, 4, and 2 ounce sizes.

plywood, particleboard, high pressure laminates, cloth, paper, and many other porous materials. It is often used to join veneers of wood.

Liquid hide glue can be used for edge or face gluing, dowelling, mortising, veneering, and various types of laminating. It can be applied by roller or dip spreader, pressurized glue cans, pressure oil cans, polyethylene plastic applicators (Fig. 2-2A), brush, or stick. It combines a rapid set with a long open or closed assembly time. Normally, liquid hide glues set (harden) in 2 to 3 hours, but require at least 8 hours. During hot, humid weather, this time may need to be increased somewhat. The speed of set can be increased by: increase of temperature; decrease of humidity; increase of air circulation; or decrease of spread (as long as it is sufficient to give a bead squeeze-out). Drying time and proper room temperature can be critical under certain circumstances. Bonding with this glue also requires clamping, but it does not creep under a load. The glue line dries to light brown or amber color. The major disadvantages of these glues are their poor water- or moisture-resistance and the importance of temperature control in their use.

If you are a purest, the flake, ground, or cake forms of hot hide glues are still available. Soak the glue in lukewarm water overnight, being sure to make it according to the manufacturer's instructions. To heat the glue, use glass ovenware or metal containers, double-boiler fashion, to keep it at about 150°F. Thermostatically controlled electric glue pots are available that keep the glue at its proper application temperature. Heat only the quantity needed; frequent reheating weakens the glue. It sets fast, but requires tight clamping for proper bonding.

Both liquid hide and hot hide glues are not affected by finishes or the finishing solvents. Hide glues do not load sander belts.

Fig. 2-2: (A) Liquid hide glue being applied from a polyethylene bottle. (B) Making a furniture repair with an aliphatic glue.

Polyvinyl Acetate (White) Glues. One of the most commonly used wood adhesives, polyvinyl acetate (PVA) glue can be used for all types of interior woodworking jobs, where waterproof joints are not required. It is also a good adhesive for paper, cloth, cardboard, leather, and other porous materials.

White glues are nontoxic, odorless, and nonflammable; children can use them with safety. They spread smoothly without running or evaporating. Clamping with moderate pressure is necessary, and workpieces should be kept at room temperature. While the adhesive sets in about 1 hour, full strength is obtained in approximately 24 hours. Surplus glue should be cleaned up with a damp cloth. When dry, it is translucent. Joints formed with polyvinyl white glue will withstand only moderate stress. It should not be used on bare metal because it causes corrosion.

Aliphatic Resin Adhesives. Possibly the fastest growing in popularity of all wood adhesives, the aliphatic resin adhesives work very much like white glue. They are, however, stronger than white glues. They have fast initial tack and set in 20 to 30 minutes. They need clamps for only half an hour and cure to full strength in 24 hours. Aliphatics offer a special advantage to owners of poorly heated workshops because they can be used at temperatures from 45° to 110°F. A high-strength bond can be had at this low temperature, but a longer curing period is required.

Aliphatics have good heat resistance and are not affected by the solvents in varnish, lacquer, or paint. Also, they can be sanded. Therefore, pieces that have been glued with aliphatics can more easily be finished using these materials than can pieces that have employed white glues, which have poor solvent resistance. An aliphatic resin adhesive can even be "dyed" or pre-colored with water soluble dyes to match the material being repaired or finished. It is good for furniture repairs (Fig. 2-2B) and can be used on cloth, paper, leather, and other porous surfaces.

Some aliphatic glues are now made that contain thixotropic features. This is desirable with some assembly glues, since the squeeze-out does not run down the joint when it is clamped.

Water-Mixed Adhesives

As their name implies, water-mixed adhesives come in a dry powder form, which must be mixed with water to make them spreadable. Casein and urea-formaldehyde glues are the water-mixed types most commonly used in the home workshop. With either material, be sure to follow the manufacturer's recommendations as to the powder-water mixing proportions. Also, keep both casein and urea glues in tightly sealed containers because even the moisture in the air can make the powders lumpy and crusty.

Casein. An old reliable wood glue made from milk protein, casein glue comes as a light beige powder which you mix with water. The glue is not waterproof, but it is moisture-resistant and has long been used on exterior jobs. Because it is nontoxic, it is often used in toy construction.

Casein glue can be applied at low temperatures (any temperature above freezing) and requires only moderate clamping pressure for 2 to 3 hours. However, the material may stain some dark or acid woods, and it tends to dull cutting tools when dry. It is excellent for use on oily woods, such as teak, lemonwood, and yew, that will not take other kinds of wood glue.

Urea-Formaldehyde (Plastic Resin) Glues. This light-colored, highly water-resistant group of wood glues is very strong, but brittle if the joint fits poorly. They will not stain acid woods, such as oak and mahogany, but should not be used with oily woods.

Plastic resin glue must be used at temperatures of 70°F or above. For maximum strength, joints must be smooth and fitted accurately. Firm clamping pressure must be maintained for at least 12 hours to insure a good bond. Plastic resin glue is resistant to rot and mold and will leave little or no glue line to mar the finished appearance. When mixed with wheat or rye flour and water in recommended proportions, it provides an easy way to glue veneers at relatively low cost. Plastic resin glues are nonflammable.

Comparison of Typical Water-Mixed and Two-Part Adhesives

	Casein	Plastic Resin	Resorcinol
Appearance	Cream	Tan	Dark reddish brown
Viscosity	35-45,000 cps	25-35,000 cps	30-40,000 cps
Speed of Set	Slow	Slow	Medium
Strength (ASTM Test)	2,800 plus psi	2,800 plus psi	2,800 plus psi
Stress Resistance	Good	Good	Good
Moisture Resistance	Good	Good	Waterproof
Heat Resistance	Good	Good	Good
Solvent Resistance	Good	Good	Good
Gap Filling Ability	Fair to good	Fair	Fair
Wet Tack	Poor	Poor	Poor
Working Temperature	32°-110°F	70°-100°F	70°-120°F
Film Clarity	Opaque	Opaque	Opaque
Film Flexibility	Tough	Brittle	Brittle
Sandability	Good	Good	Good
Storage (shelf life)	1 year	1 year	1 year

Two-Part Wood Adhesives

A two-part adhesive is one in which the glue resin and catalyst are packaged in two separate containers. To spread the adhesive, the two parts must be mixed together as specified by the manufacturer. Because most two-part adhesives must be mixed and applied fairly quickly to assure that gluing operation is completed before they begin to set, it is a good idea to mix two-part adhesives in relatively small quantities. There are two common two-part wood adhesives: resorcinol and acrylic resin adhesive. In addition to wood, the latter type bonds to almost everything—metal, glass, concrete, but not plastic. Two-part adhesives are more expensive than the water-mixed and ready-to-use wood glues.

Resorcinols. Resorcinols are completely waterproof, high-strength adhesives that are primarily intended for wood applications. They are excellent for outdoor furniture and for items immersed in water, such as boats (even toy boats). Resorcinols are a two-part adhesive, packaged in double cans. One can contains a cherry-colored liquid resin; the other contains a tan powdered hardener, or catalyst. The result of the mix is a dark brown, loose paste with a pot life of about 2 to 3 hours. When the mixture gets too thick to spread, it must be discarded.

When mixing a two-part resorcinol adhesive, use matching measuring cups and spoons (Fig. 2-3). This is to prevent any hardener from getting into the resin remaining in the original container by using the same utensils. Protect your eyes from both parts when mixing this type of adhesive. As powder has a tendency to pack down during storage, shake the *closed* powder can to fluff it and to make the measurement accurate. A stiffer mixture requires a shorter setting time than a wetter mixture.

Wood parts must be firmly clamped until the adhesive is completely dry. It cures in about 10 to 12 hours at 70°F, 7 to 9 hours at 80°, and 4 to 6 hours at 90°;

Fig.2-3: Accurate measuring is important when mixing a two-part adhesive. Mixing the activating powder into the resin.

but, the clamps should not be removed for 16 to 24 hours. It cleans off easily with warm water while it is still wet, but once it has hardened, it cannot be removed. Resorcinols have good gap filling properties and do not creep under stress loads. They withstand freezing, boiling water, heat, fungus, and mild acids and alkalis. Their major disadvantages are that they must be used at temperatures above 70°F, they have a relatively short pot life, and they require a long curing time.

Acrylic Resin Adhesive. This two-part adhesive (liquid and powder) provides an extremely strong bond that is waterproof and is not affected by gas or oil. Acrylic resin adhesives are good for filling gaps or cracks in objects that hold or are in water. It is most important that you follow the manufacturer's directions on the container to get the proper proportions for the job you are doing. Drying and setting time is controlled by the amounts mixed; for example, 3 parts of powder to 1 part of liquid will set in about 5 minutes at 70°F. Changing these proportions will allow for faster or slower drying and setting time. Acetone can be used as a cleaning solvent. The glue line color of the dried adhesive is tan.

To control penetration of the adhesive, a pre-brushing with pure liquid component is frequently used prior to applying the mixed acrylic adhesive. When bonding metals, the joints are usually made with adhesives alone. But, in either bonding situation, an acrylic type adhesive sets too fast for large area work and is generally used for heavy-duty repairs.

Two other popular wood glues—contact cement and neoprene construction adhesive—are fully described in Chapters 5, 6, and 7. Also, many of the specialty adhesives discussed in Chapter 8 can be used to bond wood; but in most cases, these glues are too expensive for general wood use.

JOINT DESIGN

The satisfaction obtained from any woodworking project comes not only from its initial appeal, but also from its continued attractiveness with use. Both its durability and its pristine appearance are important. The integrity of the glue joints is important to both of these aspects.

The term "joint" generally means the close securing or fastening together of two or more smooth, even surfaces. The construction quality of any wood project depends primarily on the quality of the joints. They must be neat, strong, and rigid to give the finished piece its necessary instant appeal and long-lasting durability. Keep in mind that destructive forces may be (1) applied from the exterior of a joint or (2) be induced by internal forces. External forces may be those applied by sitting on a chair or racking caused by pushing a case across a rough floor. Internal forces may be from shrinking of wood parts or the steady pull of poorly mated parts forced together by clamping pressure. Proper joint selection will alleviate the first type while correction for the second type may be concerned with good manufacturing practice, such as selection of wood species, uniformity of moisture content, and joint accuracy.

There is considerable latitude possible in joint design between a plain butt joint and a joint designed for maximum strength and durability. Some of these may even cause little or no change in external appearance. In other instances, considerable breadth in joint selection is possible if it does not show in the finished item. As a general rule, it is best to select the simplest joints that will do the job satisfactorily.

Success in gluing in general and joint design in particular depends on a knowledge of the structure of wood (Fig. 2-4A). The tendency of wood to shrink or swell across the grain is a major difficulty in construction work of any kind. It is the basic reason for the framing of panels or the assembly of cupboard backs in colonial and other furniture by the use of unglued, rabbeted, or matched planks;

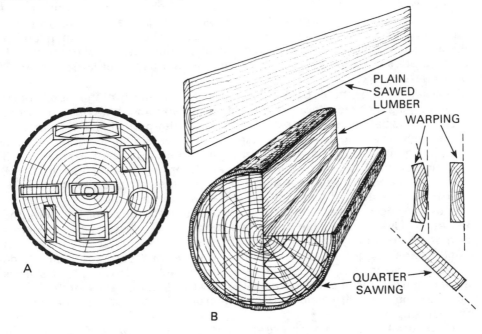

PLAIN
SAWED
LUMBER

WARPING

QUARTER
SAWING

A

B

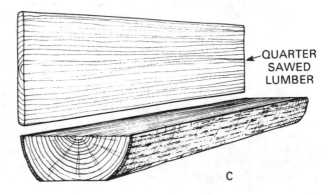

QUARTER
SAWED
LUMBER

C

Fig. 2-4: (A) Characteristic shrinkage and distortion of flats, squares, and rounds as affected by the direction of the annual rings. Tangential shrinkage is about twice as great as radial. The basic two lumber board cuts: (B) plain and (C) quarter-sawing. (Courtesy of U.S. Forest Products Laboratory.)

and it cannot be ignored when planning to make wide boards out of narrow ones. The manner in which the lumber is sawed from the log will determine its behavior. Boards are produced when bark slabs are cut off the timber to square the stick and it is ripped with parallel cuts (Fig. 2-4B). This particular method provides the greatest amount of lumber, called plain, flat, bastard, or slash cut. The outer boards have the greatest tendency to shrink due to the fact that they contain the most sap and have the flattest grain. In addition, the sap side (that lying toward the outside of log) shrinks and cups the most when drying, as a result of having the greatest amount of moisture. This tendency to shrink can be partially done away with by employing kiln-drying, a technique by which wood is artificially seasoned in sealed chambers under controlled heat and moisture.

When quarter-sawing a log, the annual rings are at right angles to the board faces (Fig. 2-4C). When radial sawing, the log is first quartered. Each quarter is held at a radial angle for all the boards. This technique produces the highest quality but the greatest waste. In quarter-sawing, warpage is greatly eliminated because since there is no heart or sap side, the lumber shrinks parallel to the faces. In addition, wood shrinks less across the rings than parallel to them. Quarter-sawn lumber is longer lasting and has a finer appearance than lumber which is slash-grained. In white oak, a type of quarter-sawn lumber, its medullary rays are split, revealing the characteristic and beautiful large, silvery or golden flakes.

An assembly joint should be designed so that the majority of the glued area is either tangential or radial grain. It should be kept in mind that with most wood species, joints between side-grain (Fig. 2-5A) and flat-grain (Fig. 2-5B) surfaces can be made as strong as the wood itself in shear parallel to the grain, tension across the grain, and cleavage. Side-grain can be held to prevent any movement by using dowels (Fig. 2-5C), splines (Fig. 2-5D), or milled surfaces (Fig. 2-5E). The tongued-and-grooved joint (Fig. 2-5F) and other shaped joints have the theoretical advantage of larger gluing surfaces than the straight joints, but in practice they do not give higher strength with most woods. Furthermore, the theoretical advantage is often lost, wholly or partly, because the shaped joints are more difficult to machine than straight, plain joints so as to obtain a perfect fit of the parts. Tongue-and-groove joints are, however, frequently used on side panels or in raised panel doors. If the panel insert is allowed to float, that is, if it is smaller than the encompassing frame, there will be less tendency for the expansion and contraction (from moisture changes) of the panel to stress the joints in the enclosing frame.

It is practically impossible to make end-butt joints (Fig. 2-6A) sufficiently strong or permanent to meet the normal requirements of furniture jointing. With the most careful gluing possible, not more than about 25% of the tensile strength of the wood parallel with the grain can be obtained in butt joints. In order to approximate the tensile strength of certain species, a scarf, serrated, or other form of joint that approaches a side-grain surface must be used. The plain scarf (Fig. 2-6B) is perhaps the easiest to glue and entails fewer machining difficulties than the many-angle forms.

It is also difficult to obtain the proper joint strength in end-to-side joints (Fig. 2-6C), which are further subjected in use to unusually severe stresses as a result of unequal dimensional changes in the two members of the joint as their moisture content changes. It is therefore necessary to use irregular shapes of joints, dowels (Fig. 2-6D), tenons, glue blocks (Fig. 2-6E), or other devices to reinforce such a joint in order to bring side grain into contact with side grain or to secure larger gluing surfaces.

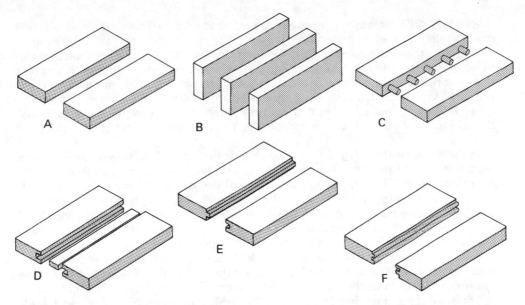

Fig. 2-5: Edge glued and face glued joints are strong, but they can be made stronger by using dowels, splines, and milled surfaces.

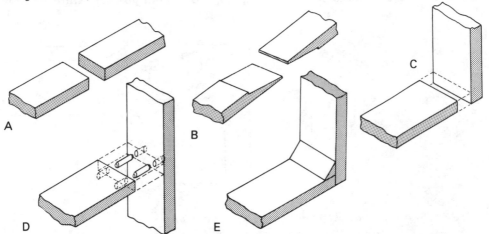

Fig.2-6: End-to-end and end-to-side butt joints and methods of reinforcing them.

While there are well over 100 kinds of joints used in woodworking, they can be grouped into the following several basic types.

Butt Joints

The butt joint is the simplest of all joints. Though it is extremely simple to make, the edges to be joined must be tested for *absolute* squareness before the pieces are fitted together. But, even then, as already mentioned, the plain butt end grain to end grain or plain butt end grain to side grain joints have little strength, unless the joints are strengthened by either dowels or wood glue blocks.

Doweling. Dowel rods are made of hardwood—generally maple or birch— and are sold in 36″ lengths and diameters from 1/8″ to 3″, including the common 1-1/2″ clothes pole dowel. The sizes most often used for cabinetwork are

3/8" and 1/2" diameters. The dowel's hardness and the fact that it is used with grain at right angles to materials joined are the reasons for this added strength.

Actually, there are two methods used in doweling, the open method and the blind method. In the open method (Fig. 2-7A), a hole is drilled completely through one piece of wood and deeply into or through the piece to be joined. The dowel is coated with glue and pushed completely through the drilled holes, joining the pieces. The remainder is then sawed off flush with the outer surface. Thus, dowel stock for open doweling is kept long to allow for a flush cut after the joint is made.

In the blind method (Fig. 2-7B), holes are drilled part of the way into each piece from the joined faces. A rule of thumb is to drill the holes in each piece to a depth of approximately four times the diameter of the dowel. A dowel is then glue-coated and inserted in one hole, and the second piece is pressed onto the protruding dowel end. The length of the dowel rod should always be cut about 1/8" to 1/4" shorter than the total of the two holes. Chamfer the dowel at each end to aid in location during assembly. The fit of the dowel should not be so tight that the glue is all pushed to the bottom of the hole. Optimum size is a snug fit, such that the dowel, spline, or tenon can be pushed in with a finger but not loose enough to wobble in the hole. A 1/64" clearance is a good fit to obtain.

Consistency of dowel and hole diameter is difficult to obtain. Holes and dowels should be checked frequently for uniformity of diameter. Gluing should be done soon after machining to prevent change of size by change of moisture content.

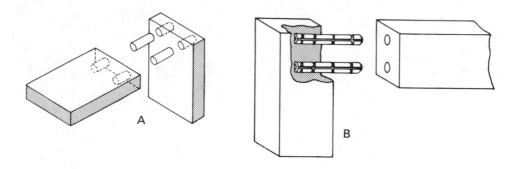

Fig.2-7: (A) An open dowel joint; (B) a blind dowel joint.

Various methods are used for laying out the hole positions. A simple but accurate method is shown in Fig. 2-8A. Ordinary pins are stuck in a block of wood and placed in such a position as to come between joining members, as shown. When the joining members are pushed forcibly together, the pinheads make an impression on each piece, which serves as a guide when drilling the dowel holes.

Double-point thumb tacks can be used in the same manner. The more standard method of working, however, is to use dowel centers or "pops," as shown in Figs. 2-8B and C. Here, after drilling the holes for the dowels in one piece of wood, you insert dowel centers in these holes. Then, you align the two pieces of wood as they will be joined. When you press them together, the points on the dowel centers mark the second piece of wood. It is now possible to drill holes at these center marks. When the pieces are connected with dowels, the blind dowel joint is perfectly aligned. Dowel centers come in assorted sizes to fit holes from 1/8" to 1" in diameter.

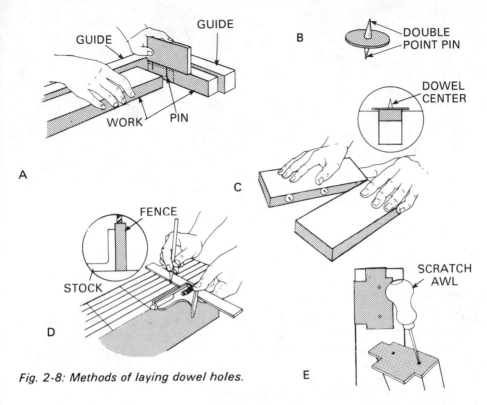

Fig. 2-8: Methods of laying dowel holes.

When locating dowel holes in a series of boards that will be joined edge-to-edge, position the board edges and butt them surface to surface. Using a combination square (Fig. 2-8D), mark the hole location on one edge and carry the line across all the pieces. Identify the board faces which will be at the top after assembly. Drill dowel holes in the edges at the cross lines, using a dowel jig or drill guide such as the one shown in Fig. 2-9, that will gauge holes automatically.

When doweling joints other than those at edges, it is a good idea to make a template of stiff cardboard, thin plywood, hardboard, or even sheet metal if the long-term use justifies it. Drill 1/16" or smaller dowel center holes at the desired locations. Locate the template accurately first on one piece, then the other, to mark dowel centers with a center punch or awl (Fig. 2-8E).

When using the open method of doweling, the problem of aligning drilled holes does not occur. Even when poorly centered, the holes match and the dowel can be driven through both holes. In the blind system, however, two separate holes must be drilled—and here trouble can develop unless a jig is used.

In some instances, the drilling device is clamped into position and the wood is guided to it along measured channels to ensure proper centering of the holes. In others, the wood is clamped and the drilling device is guided along identical channels for all holes drilled. In either case, the guide is actually a jig. Furthermore, it is also possible to arrange the jig to control depth of hole drilled.

Since most of the strength of an assembly joint comes from side or face grain, a complete spread of glue on the sides of the dowel hole is desirable. If the glue is to be spread either on the dowel or in the hole, spreading glue in the hole is usually more desirable. Spreading glue on both tenon or dowel and the hole is better than spreading on only one.

Fig. 2-9: How a dowel jig is set up.

Some glue spreaders put glue in the bottom of the hole and hope that when the dowel reaches the bottom of the hole, the glue will be forced up around the dowel. If this does not occur, a weak joint will result. To be successful, all of the following conditions must be fulfilled:

1. Adequate glue must be spread.
2. The dowel must go to the bottom of the hole.
3. The dowel must fit loosely enough to allow the glue to come up around it.

Flutes or spiral groove dowels (Fig. 2-10) will allow the glue to come up in the grooves but does not assure the presence of glue outside the grooves. Figure 2-11 illustrates metal tips which can be made and adapted for use with applicator bottles to spread glue on the sides of grooves and dowel holes. A flange must be brazed or soldered on the lip. This flange fits under the plastic cap.

Glue or Corner Blocks. Glue blocks are small square or triangular pieces of wood used to strengthen and support the two adjoining surfaces of a butt joint. Remember that while this joint reinforcement features a strong face-to-face gluing surface, varying moisture contents will cause a differential wood movement, since the grain direction of the block and the substrate are at right angles to each other. This will highly stress the joint. Because of this, a number of short blocks is preferable to one long one.

Mortise-and-Tenon Joints

The mortise-and-tenon is a very good joint, stronger and more widely used than the butt joint. It is one of the best techniques used in fine furniture making. All enclosed mortises (those with material on four sides) should be cut with a mortising attachment on a drill press or with a router.

The tenon for the mortise-and-tenon joint can be cut in a number of different ways (Fig. 2-12), depending on the equipment available and the nature of the joint. In any case, thickness of the tenon should not exceed one-third the thickness of the mortised member. Some of the more popular variations of the simple mortise-and-tenon (Fig. 2-13) are described below.

The *haunched tenon* is employed where added tenon strength is needed and where partial exposure on top is not objectionable.

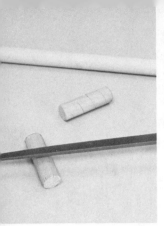

A B C

Fig. 2-10: Three methods of fluting or spiralling dowels: (A) with a triangular file; (B) with pliers, or (C) with a band saw.

Fig. 2-11: Metal tips make glue spreading on dowel surfaces easy.

Fig. 2-12: (A) Cutting a tenon on a table saw; (B) cutting a mortise on a drill press.

A B

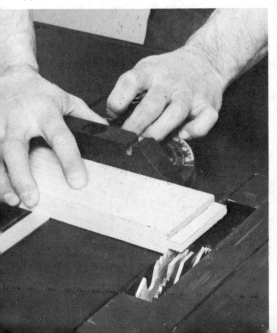

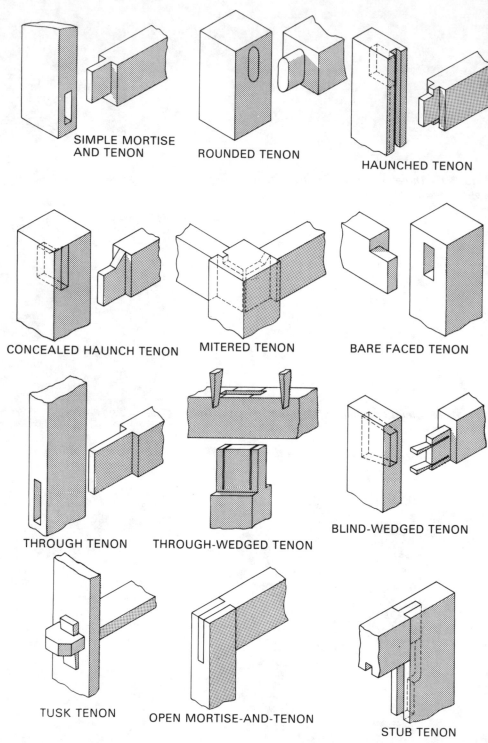

SIMPLE MORTISE AND TENON

ROUNDED TENON

HAUNCHED TENON

CONCEALED HAUNCH TENON

MITERED TENON

BARE FACED TENON

THROUGH TENON

THROUGH-WEDGED TENON

BLIND-WEDGED TENON

TUSK TENON

OPEN MORTISE-AND-TENON

STUB TENON

Fig. 2-13: Common mortise-and-tenon joints.

The *concealed haunched tenon* gives the needed extra strength to the joint without showing a break at the end.

The *mitered tenon,* frequently utilized in table construction, is used to secure the maximum length of tenon. Each joint is a simple mortise-and-tenon with the tenon end mitered at 45°, as shown. The two mortises meet at 90° inside the vertical (leg) member. Mitering of the tenon ends allows for deeper tenons.

The *bare-faced tenon* has but one shoulder and it is used when a tenoned piece is thinner than a mortised piece.

A *through tenon* is a useful joint in some types of furniture. Where added strength and resistance to pulling apart are required, wedges can be used. The two ends of the mortise are sloped outward to provide room for the wedges, which are about half the tenon in length.

The *blind-wedged tenon* is used in the same way as the through-wedged tenon but can be employed in locations where the through-wedged cannot.

The *tusk tenon* that goes right through and is locked with a key, peg, or wedge goes back to medieval days. They were more common in some middle European countries than in Great Britain, but they may be found in some Colonial furniture. The use of a wedged or keyed tenon would improve an ordinary tenon, where the fit is not good.

The *open mortise-and-tenon* is most commonly used in simple frame construction where an exposed tenon end is not objectionable.

Although not a true mortise-and-tenon joint, the *stub tenon* is sometimes used in frame construction and is made with a short tenon that fits into the groove of the frame.

Dovetail Joints

The interlocking of two pieces of wood by special fan-shaped cutting is called a dovetail joint. It is used extensively in making fine furniture, drawers, and in projects where good appearance and strength are desired. A dovetail joint has considerable strength because of the flare of the projections, technically known as pins, on the ends of the boards, which fit exactly into similarly shaped dovetails.

There are four basic dovetail joints (Fig. 2-14):

1. Through or lap dovetails are joints where the dovetail and pin are both clearly visible from two sides of the joint.

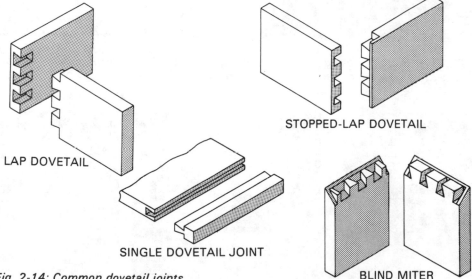

STOPPED-LAP DOVETAIL

LAP DOVETAIL

SINGLE DOVETAIL JOINT

BLIND MITER

Fig. 2-14: Common dovetail joints.

2. Stopped lap dovetails or half-blind dovetails are joints where the face of one board is perfectly smooth and the pins and dovetails are visible on the face of the other.

3. Blind dovetail joints are those where all the cutting is done without marring the outside face of either of the two pieces.

4. A single dovetail (Fig. 2-15), as its name implies, is just one dovetail cut into a corresponding frame. It is useful in making a table base in which multiple legs are attached to a central post.

Dovetails can be cut by hand using a dovetail saw and small chisel or with a special template or pattern with a drill press or router (Fig. 2-16). The latter two methods are highly accurate and considerably faster than cutting by hand.

Dado Joints

A dado joint is formed when one piece of wood is set into a groove or dado cut into another. There are many variations of a dado joint (Fig. 2-17) used in cabinetwork and furniture making. For instance, a standard or housed dado joint is a groove that is cut in one piece of wood to the exact thickness of the second piece to be joined. Sometimes a dado is stopped on one or both sides.

The shouldered dado is generally employed for drawer joints, but is also used extensively in shelf construction. The full-dovetail dado is made by first cutting a

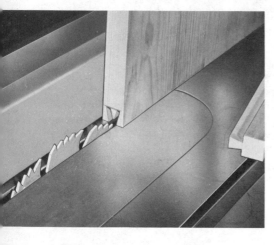

Fig. 2-15: Cutting a single dovetail on table saw.

Fig. 2-16: Steps in making a dovetail joint with a router: (A) Putting the template in place; (B) cutting with the router; and (C) the results.

A   B   C

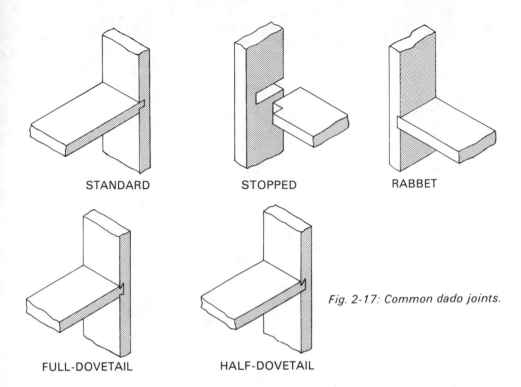

STANDARD STOPPED RABBET

FULL-DOVETAIL HALF-DOVETAIL

Fig. 2-17: Common dado joints.

mortise or slot. Then, cut the dado to the narrowest width. The half-dovetail dado is cut the same way except that the angle cuts are made on only one side of the joint. Actually, a dado is a poor joint for gluing as it has the same deficiency as an end-to-side grain joint—low glue joint strength. The dovetail dado does have more strength than a plain dado, as the shear strength of the wood prevents the rupture of the joint.

Lap Joints

Bring two workpieces together, notch them equally where they overlap, and you have either a cross- or middle-joint, a half-lap joint (end-to-end, at right angles), or a tee half-lap joint (end-to-side, at right angles). The desired visual effect of such joints is that the two thicknesses overlap within a single thickness. A well-made lap joint is a strong joint in that the glued surface is a face-to-face surface (Fig. 2-18).

The end or corner-lap joint is made by halving and shouldering the opposite sides of the ends of the workpieces and then joining them at right angles. When one of the cuts is made at a point other than the end of the workpiece, you have a center, or middle, lap. This joint is used most commonly when joining a rail to an upright.

The edgewise cross-lap or middle lap is another joint that is made by halving the two pieces at right angles, but in this case the notches are cut in the edges rather than the surfaces of the two members. This is particularly useful when making framing or partitioning, and often both cuts can be made at the same time.

The full-dovetail half lap and the half-dovetail half lap are fancier versions of the corner and cross lap which are often seen in old cabinetwork. In both these joints the dovetail is half-lapped into a rail, crosspiece, or upright, and the two

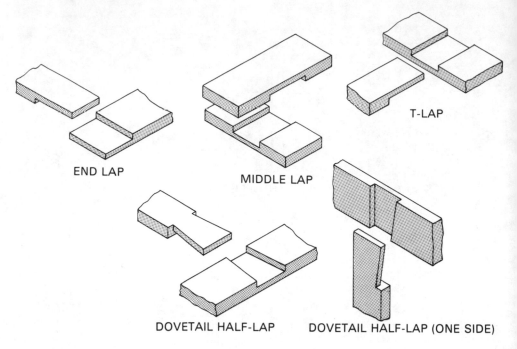

Fig. 2-18: Common lap joints.

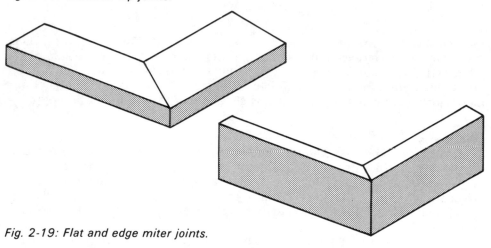

Fig. 2-19: Flat and edge miter joints.

pieces are joined at right angles. The main difference between the two is that the half-dovetail lap is not as difficult to make.

Miter Joints

The miter joint is primarily for show. For example, it may be used for an uninterrupted wood grain around edges (side to top to side of a cabinet) or at corners (a picture frame). The joining ends or edges are usually cut at angles of 45°, then glued and clamped (Fig. 2-19). The 45° glued miter joint is stronger than the end-to-end or end-to-side joint, but not nearly as strong as a face-to-face glued joint. For a high quality mitered joint, various modifications are possible which contribute to the strength of this construction. These include splines, keys, feathers, or dowels (Fig. 2-20).

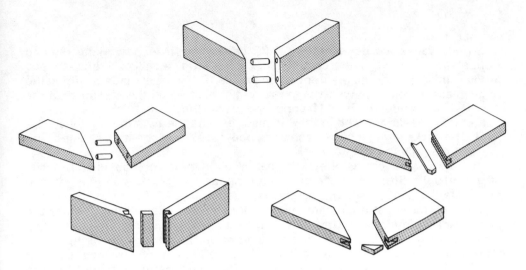

Fig. 2-20: Flat and edge miters with dowels, splines, and feathers.

Splines are used to strengthen all types of joints from plain butt to fancy miters. The spline itself is a thin strip of hardwood or plywood inserted in a groove
cut in the two adjoining surfaces of a joint. The groove is cut with a saw blade or
dado head to a specific width and depth. (The groove for the spline is commonly
run in with the dado head, 1/4" being usual for 3/4" to 1" stock, although a 1/8"
spline, a single saw cut, is sometimes used, especially for miters.) A thin piece of
stock is then cut to fit into this groove. The spline stock should be cut so that the
grain runs at right angles to the grain of the joint.

A very simple way to produce splines is to cut up scrap pieces of 1/8" plywood
or hardboard. A supply of these can be kept on hand. The advantages of the plywood is its strength in each direction and its constant thickness. Quite probably
your saw blade cuts a 1/8" kerf which is just right.

A key is a small piece of wood inserted in a joint to hold it firmly together. The
key is sometimes called a *feather*. It is often placed across the corners of miter
joints.

Lapped miter joints (Fig. 2-21) are generally used only if it is desirable to have
the miter show on one side. This joint will give the appearance of a conventional miter from one side, yet will be considerably stronger because of the increase in the area of contacting surfaces.

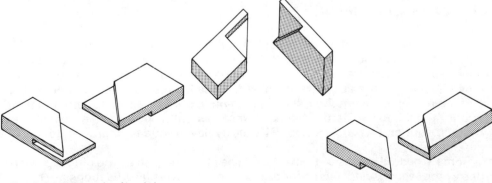

Fig. 2-21: Lapped miter joints.

Corner Joints

Various types of corner joints (Fig. 2-22) can be used in woodworking, ranging from a simple rabbet to the complex lock miter joint. A rabbet is an L-shaped groove cut across the edge or end of one piece. The joint is made by fitting the other piece into it. The width of the rabbet should equal the thickness of the material, and its depth should be one-half to two-thirds the thickness. The rabbet joint conceals one end grain and also reduces the twisting tendency of a joint. The backs of most cases, cabinets, bookcases, and chests are joined with the end grain facing the back.

The lock-corner joint is one of the better joints to use on chests and special boxes. Allow a little tolerance between tongues and grooves so that you can assemble the joint by sliding the pieces together.

Although the illustrations and explanations in this chapter might make the joints appear simple, it is extremely easy to spoil expensive material by making such common mistakes as cutting through or on the wrong side of the layout lines, not properly identifying mating surfaces, and/or not following the proper sequence of steps. Therefore, when attempting a joint for the first time, it is wise to practice on scrap material of the same size to check settings, fits, and final results. Sometimes it might be necessary to make the same joint several times before obtaining satisfactory results. Remember, when woodworking, "practice makes perfect."

CONDITIONING AND PREPARING THE WOOD FOR GLUING

Before any wood project can be glued, all parts must be sanded to their desired smoothness, and they should accurately fit together. In other words, all parts should be considered to be in a "finished" condition when they are glued and clamped.

To get glued assemblies of maximum quality, joints should be cut accurately with sharp tools. A joint should fit snugly when assembled and/or clamped. Glue should not be expected to act as a gap filler although some glues do fill

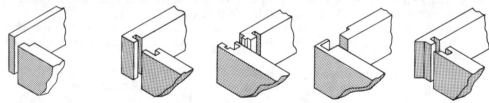

RABBET JOINT   MILLED CORNER   LOCK JOINT   BOX CORNER   LOCK MITER JOINT

Fig. 2-22: Common corner joints.

minor irregularities better than others. A dull cutting tool may glaze or even burn a joint. When this occurs, the surface fibers are beaten down, closing them to adequate glue penetration. A dull saw or drill bit could also loosen but not remove fibers from the surface to be glued. Since the glue can only adhere to the surface presented, even though the adhesion to the fibers is excellent, the joint strength will be low, as the fibers themselves will rupture easily. Therefore, keeping your cutting tools sharp will quickly pay off due to improved quality of glued joints.

Some woods glue easier than others. The chart here gives the gluing properties of the woods widely used for glued products. The classifications are based on the average quality of side-grain joints of wood that is approximately average in density for the species, when glued with wood glues. A species is considered to be glued satisfactorily when the strength of the joint is approximately equal to the strength of the wood. Whether it will be easy or difficult to obtain a satisfactory joint depends upon the density of the wood, the structure of the wood, the presence of extractives or infiltrated materials in the wood, and the kind of glue. In general, heavy woods are more difficult to glue than lightweight woods, hardwoods are more difficult to glue than softwoods, and heartwood is more difficult than sapwood.

HARDWOODS

Group 1	Group 2	Group 3	Group 4
(Glue very easily with different glues under wide range of gluing conditions.)	(Glue well with different glues under a moderately wide range of gluing conditions.)	(Glue satisfactorily under well-controlled gluing conditions.)	(Require very close control of gluing conditions or special treatment to obtain best results.)
Aspen. Chestnut, American. Cottonwood. Willow, black. Yellow-poplar.	Alder, red. Basswood. Butternut. Elm: American. Rock. Hackberry. Magnolia. Mahogany. Sweetgum.	Ash, white. Cherry, black. Dogwood. Maple, soft. Oak: Red-. White. Pecan. Sycamore. Tupelo: Black. Water. Walnut, black.	Beech, American. Birch, sweet and yellow. Hickory. Maple, hard. Osage-orange. Persimmon.

SOFTWOODS

Group 1	Group 2	Group 3
Baldcypress. Cedar, western red-. Fir, white. Larch, western. Redwood. Spruce, Sitka.	Cedar, eastern red-. Douglas-fir. Hemlock, western. Pine: Eastern white. Southern yellow. Ponderosa.	Cedar, Alaska-.

The condition of the wood must be taken into consideration before attempting any gluing operation. For instance, the amount of moisture in wood affects both the rate at which glue dries and the strength of the finished joint. If wood has too much moisture before it is glued, a weak joint results. If wood takes on too much moisture during gluing, it first swells and then shrinks. This can set up stresses along the glue line which could result in failure of either the wood or the glue. Thus, for a successfully glued joint, the glue must be rigid to resist the stresses applied to the glue line by moisture change, particularly before this moisture interchange is slowed by the finish. Since wood absorbs and releases moisture faster through the end grain than through the radial or tangential face, a moisture change sets up stresses in the panel. In a dry atmosphere, the moisture leaving through the end grain tends to shrink the end of the panel. If the glue joint is weaker than the wood, it will open; otherwise, when the stress becomes great enough, the wood may split. Many times inferior quality joints will remain closed until subjected to a moisture change, when the stress will break the joint instead of the wood. All gluing should be done when the wood is at a moisture content below 10%. However, some woods glue better at even lower moisture content.

Failure to take the moisture content of the wood into account before gluing can cause a "sunken" joint. For example, when boards are glued edge-to-edge, the wood at the joint absorbs water from the glue and swells. If the glued assembly is surfaced before this excess moisture is dried out or distributed, more wood is removed along the swollen joints than elsewhere. Later, when the moisture throughout the panel is equalized, the swollen wood in the vicinity of the glue line will shrink more than the rest of the panel, leaving the wood near the glue line depressed below the rest of the surface (Fig. 2-23). Sunken joints are more obvious with glossy finishes than with dull or matte finishes.

Sunken joints can be corrected by: (1) seasoning the glue joint longer before machining; (2) using less glue so that only a slight bead squeeze-out occurs on clamping; (3) using methods to get faster drying at the glue line; or (4) the use of faster drying glues.

If the boards in an edge-to-edge panel are not of uniform moisture content, after gluing and machining to a uniform thickness, the higher moisture content boards will shrink more than the lower moisture content ones, giving the finished panel a rippled effect at each glue line. The same effect will occur if plain-sawn and quarter-sawn boards of the same moisture content are placed next to each other and the whole panel changes markedly in moisture content. The quarter-sawn board will shrink in thickness more than the plain-sawn board (Fig. 2-4A). Plain-sawn lumber shrinks more on the sap side than on the heart side. Therefore, if panels consist of very wide boards or if several boards are glued together with the sap side facing one surface, the panel will warp markedly toward the sap side if the moisture content is lowered. If no board going into the panel is over 4" wide and the annular ring direction is reversed on adjoining boards, the cupping or warping will alternate and be less noticeable.

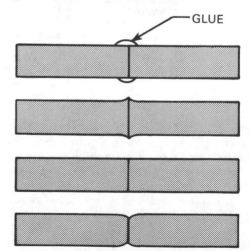

GLUE

Fig. 2-23: This is what will happen if a glued-up panel is surfaced while there is a raised glue line.

Gluing and Clamping 3

Any good wood gluing depends on several factors, the most important of which are:

1. Selection and preparation of the adhesive;
2. Preparation of the wood surfaces to be joined;
3. Design of the joints;
4. Application of the adhesive;
5. Assembly of the parts and proper use of clamps and other clamping devices;
6. Post treatment or exposure conditions of the glued joint. The first three of the six steps to good wood gluing were fully covered in the previous chapter; the last three are covered in this chapter. But, before taking a look at these, let us first look at the items needed for a good gluing job.

CLAMPING DEVICES

Although some glues and adhesives will form a reasonably strong bond when set at mere contact pressure, others require clamping. Actually, the only purpose for clamping is to hold the parts in intimate contact until the adhesive gains enough strength to do so. Select the design and strength of the clamping device to bring the parts tightly together. Among the clamps which the craftsman might use for shop gluing jobs are the following:

Handscrew or Parallel Wood Clamps. These clamps are the oldest type of woodworking shop holding devices known. They consist of two parallel hardwood jaws connected with two opposite working screws. The handle end of the screw runs freely in the hole, while the other end works in a threaded hole in the other jaw. The other screw works the same, but in the opposite direction. Handscrews are a generally preferred holding device for nearly all types of shop projects and repair work. They grip and hold odd shapes securely and will not mar highly finished surfaces. Handscrew clamps are made in sizes from 5/0, which has a jaw length of 4" and a maximum opening between jaws of 2", to size 7, with a length of 24" and a maximum opening between jaws of 17". If using *new* clamps that have oil-finished jaws, clamp them tightly against blotting or other absorbent paper on a block before using them on actual work.

To use handscrew clamps correctly, learn the habit of grasping the "end" spindle (Fig. 3-1A) with the right hand; then the direction for "swinging" or rotating the handscrew to open the jaws will always be the same. Rapid adjustment of the handscrew is obtained by proper swinging. Hold the handles firmly, arms extended, and, with a motion of the wrists only, make the jaws revolve around the spindles. When the jaw opening is approximately correct, place the handscrew on the work with the "end" spindle either to your right or in the upper position, and with the "middle" spindle as close to the work as possible. Adjust either or both handles so that the jaws grip the work easily and are slightly more open at the end. Turn the end spindle clockwise to close the jaws onto the work (Fig. 3-1B). Final pressure is applied only by means of the end spindle. The middle

A   B

Fig. 3-1: (A) Proper way to open or close a handscrew clamp. (B) The jaws should be parallel when the workpiece is clamped.

spindle acts as a fulcrum. Make certain that pressure is applied all along the entire length of the jaws, not just at the end or at the edge of the work. That is, in order for a handscrew clamp to be effective, the jaws must be parallel when the workpiece is clamped.

Handscrews can be used not only for gluing materials face-to-face, but also for many other clamping tasks (Fig. 3-2). For instance, if you wish to clamp a table frame to the underside of the top and hold it in place while the glue on the glue blocks is drying (or to otherwise strengthen the joint), place a stiff piece of material across and under the frame. Clamp this to the top, providing, of course, that the clamp jaws are not long enough to reach into the frame.

Fig. 3-2: Handscrew clamps are valuable in furniture repair work. (Left) A large clamp holds two small handscrews when the adhesive cures.

C- or Carriagemakers Clamps. Named as such because their bodies are shaped like the letter C, these clamps are the ones most commonly used in the workshop. They consist of a steel frame threaded to receive an operating screw with a swivel head. C-clamps are made for light, medium, and heavy service with a variety of openings from 5/8" to 12". (Sizes are based on the maximum opening between jaws.) The depth to the back of the clamp usually ranges from 1" to approximately 4", depending on the size of the clamp opening. Deep-throated styles are available which have greater depth.

C-clamps are adaptable to many uses some of which include holding edging on veneer strips, clamping miter corners (Fig. 3-3), as well as many "spot" clamping jobs. The ball joint at the foot of most C-clamps is designed to swivel so that work which is not absolutely flat can still be securely clamped. It is a good idea to always purchase C-clamps (and most clamps for that matter) in pairs of the same size, as that is probably the way you will use them.

Since bare metal against wood can leave undesirable holes or marks, use small pieces of scrap wood beneath the metal jaw pads (Fig. 3-4). These pro-

28

Fig. 3-3: C-clamps with the proper shaped wood blocks can be used to hold corners together while the glue sets.

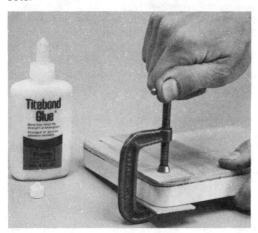

Fig. 3-4: When using C-clamps, be sure that the jaws are protected by wood blocks.

tective wood pieces also serve to uniformly distribute the clamp's pressure. Thus, when using a C-clamp, open the jaws just far enough to fit in the workpiece plus the protective wood pieces; you will then be able to quickly put pressure on the joint before the adhesive begins to set. Tighten by hand only; tightening by mechanical means (pliers) could damage either the clamp or the workpiece.

Bar Clamps. These clamps, often called cabinet or furniture clamps, are generally used for long-span work, such as across a table top or across the front of a chair. They consist of a steel bar and two clamp jaws (Fig. 3-5). One jaw, on the end of the bar, is fixed in position. The other moves the length of the bar as needed. Actually, there are three styles of bar clamps: *sliding head* style in which the screw adjustment portion is the movable jaw; *fixed head* style in which the screw portion is the stationary jaw; and *hinged head* style. The latter has a swivel plate attached to the stationary head which permits more versatility in the mounting of the bar clamp. The normal maximum opening of standard bar clamps ranges from 2' to 8'. Smaller "quick-set" bar clamps are available with openings of 6" to 30".

When using any bar clamp, place the fixed clamping jaw against one side of the workpiece and slide the movable jaw against the other side; then tighten the clamp with the screw handle by hand. Be careful not to over-press. Also, scrap wood pads should be employed under the jaws to prevent marring.

Pipe Clamps. Pipe clamps operate in the same manner as bar clamps except that they slide on pipe. These clamps (Fig. 3-6) are available to fit either 1/2" or 3/4" diameter iron pipe. Only one end of the pipe need be threaded. The craftsman should have several different lengths of pipe to use with the clamps. While you can use a long pipe for all jobs, the excess pipe might get in your way. The pipe, of course, must be straight and smooth.

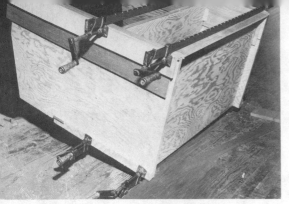

Fig. 3-5: Steel bar clamps have many uses around the shop: (left) to hold large glued cabinets together during setting time, or (right) for clamping edge-to-edge glued-up board. In the latter example, note that clamps are reversed to equalize the pressure.

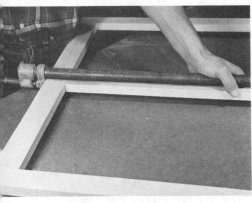

Fig. 3-6: When using pipe clamps on woods such as oak, mahogany, cherry, cedar, or walnut, do not allow the pipe to rest on the workpiece. Because of the acid pH factor of some adhesives, any contact between the glued piece and pipe clamp could cause staining. Use an air gap, thin wedges, or waxed paper to keep the squeeze-out from contacting the iron pipe.

Wood Bar Clamps. These are similar to steel bar and pipe clamps except that the sliding bar is of wood rather than metal. There are instances in gluing fine cabinetwork where the wood bar has an advantage over the metal one in that it is less likely to mar a finished surface (Fig. 3-7).

Spring Clamps. Resembling large clothespins, these hand clamps are available in sizes that open from 3/4" to 4". They are intended for lightweight clamping tasks, holding small glued-up parts and the like. More even pressure can be directed to the work area by inserting a wood block under the jaws of the clamp (Fig. 3-8). Most spring clamps have plastic-covered tips to minimize marring of the work.

Fig. 3-7: Using wood bar clamps to glue up a furniture repair.

Fig. 3-8: Spring clamps are useful lightweight clamps. To achieve more even pressure directly on the work area, insert a wood block under the jaws of the clamp.

You can make your spring-type clamps from clip-type clothespins (Fig. 3-9). To do so, while the pieces are still in rectangular form, drill the fulcrum dowel portion with a 3/8" drill at the center of the piece. Next, cut tapers as indicated in the drawing, and then drill the hold notches with a 3/8" and 3/16" drill. Break all sharp corners with garnet paper and apply two coats of varnish. A rubber band is used on either end of the clamp.

Pliers may also become an improvised spring clamp for odd pieces like round knobs and other similar shapes (Fig. 3-10). After the pliers have been squeezed on the work, tie down the handles with twine or a rubber band. If necessary, the jaws can be padded with tape.

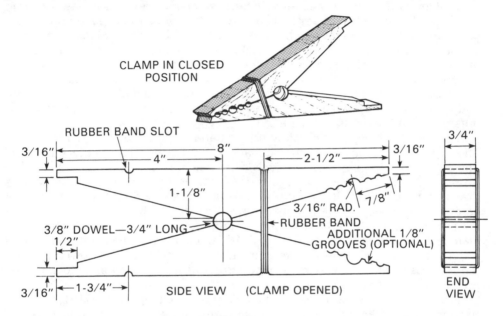

Fig. 3-9: Sketch of how to convert clip-type clothespins into spring-type clamps.

Fig. 3-10: (Left) To make slip-joint pliers into lightweight clamps, use a rubber band or a piece of string to hold the jaws. (Right) Vise-grip pliers also make good clamps as long as their jaws are padded with a wooden block, tape, or make-up pads to prevent marring the stock.

Fig. 3-11: A web clamp (left) and steel band clamp (right) are excellent for holding odd shapes for gluing.

Web or Band Clamps. These clamps solve the knotty problem of clamping round or irregular shapes where uniform pressure is required simultaneously at several joints. They are especially efficient for clamping furniture, as shown in Fig. 3-11. The canvas or steel band encircles the work and is pulled tight from either end through a screw-clamp device. The self-lock cams of the clamp hold the band securely without slippage while final screw pressure is applied. Slight pressure on the cam extensions releases the band instantly. The canvas band, which is usually about 2" wide, is recommended for most applications. To use this type of clamp, place the band around the work and pull snug at the clamp body. Then, tighten the band to the desired pressure by a crank or ratchet mechanism, depending on the make and type.

Web clamps are lightweight, low-priced band clamps with innumerable uses. The 1" wide nylon band (which is usually 12' to 15' long) can be placed around any regular or irregular shape to apply clamping pressure all around the work—drum tables, chair frames, picture frames, etc. With the band so located, it is drawn snug by hand, and final pressure is applied by means of a wrench or screwdriver applied to the hexhead slotted tightening bolt.

Edge Clamps. These handy clamp attachments are designed for steel bar clamps that have a bar not more than 5/16" thick. They provide pressure at right angles to the axis of the bar of the clamp used. Their clamping action can be employed for applying molding to the edges, drawing joints together, or for applying pressure to the middle of a broad area (Fig. 3-12).

Miter Clamps. There are several types of miter clamps (Fig. 3-13); most are used in picture frame making. The miter clamp shown in Fig. 3-14 is designed for mitering flat casing where a 5/8" diameter blind hole can be bored in the back of each piece. It forces the two ends being mitered directly against each other, no matter what the angle, with no tendency for the ends to creep along or away from each other.

The corner clamp is a flat triangular device used to clamp the corners of furniture frames, picture frames, and the like. The frame is set into the clamp, and a diagonal bolt then pushes it against the corner. In a variation of this device, pressure is exerted by two bolts which tighten against the outside corners of the frame.

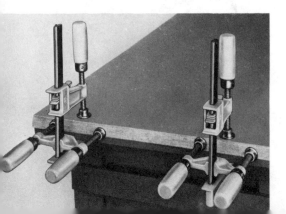

Fig. 3-12: Using an edge clamp fixture to hold a piece of molding on the edge of a table top.

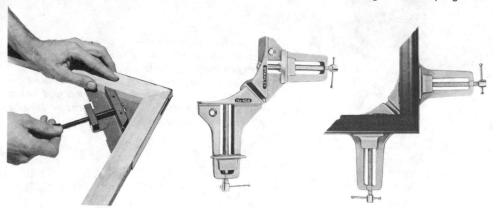

Fig. 3-13: Popular types of miter clamps.

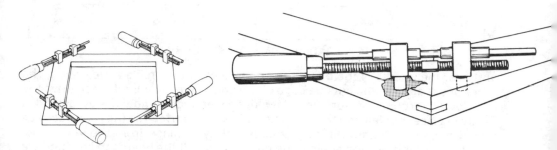

Fig. 3-14: A popular clamp designed especially for mitering flat casings.

The miter clamp shown in Fig. 3-15 is an interesting "one-evening" project that will provide you with an uncomplicated clamping jig with many advantages for mitering picture frames and similar objects. It is adjustable to any size of frame; it applies uniform pressure to all four joints simultaneously; it leaves joints visible so you can be sure they are straight and tight; it is light and easy to handle, minimizing the danger of damage to new frames or those precious old ones in need of restoration; and it eliminates the necessity of buying a separate clamp for each joint. This practical jig overcomes the disadvantages of most other miter clamps which hold work of a limited size range and apply little, if any, pressure to the joint itself.

Wedge Clamps. It is possible to make your own clamping devices to hold work during adhesive curing. For example, a homemade wedge clamp, shown in Fig. 3-16, is excellent for bonding boards that are being glued together, edge-to-edge. This simple clamp can be made from scrap lumber 3/4" to 1" thickness, 6" to 8" wide, and about one-half as long as the boards to be glued. On the wide side of the clamp material, mark each end at the midpoint from the sides. Now measure up and mark a point 1/2" above the center on one end and on the other end 1/2" below. Connect these two points by a diagonal line and saw through the material lengthwise along the line. This will give you two wedge-shaped pieces of the required angle, regardless of the length of the material.

To use the clamp, drive a nail that is about 1-1/2 times longer than the thickness of the wedges into, but not through, one of the wedges only, on the flat surface near each end. Put the wedges together as sawed, having on the outside

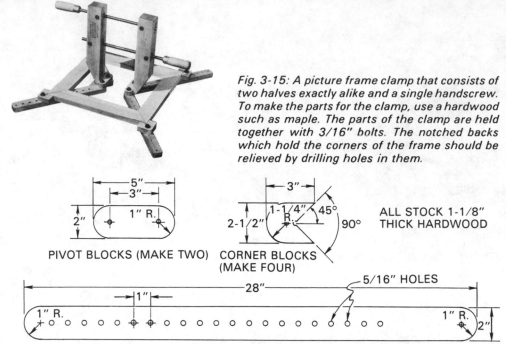

Fig. 3-15: A picture frame clamp that consists of two halves exactly alike and a single handscrew. To make the parts for the clamp, use a hardwood such as maple. The parts of the clamp are held together with 3/16" bolts. The notched backs which hold the corners of the frame should be relieved by drilling holes in them.

PIVOT BLOCKS (MAKE TWO)

CORNER BLOCKS (MAKE FOUR)

ALL STOCK 1-1/8" THICK HARDWOOD

RADIAL ARMS (MAKE FOUR)

and near the middle of the board the one into which the nails were driven. The other one is against the outside edge of the board to be clamped. Then drive the nails partly through the surface below, deeply enough to hold the outer wedge firmly in place, but leaving the nail heads above the surface so they can easily be pulled out later. Pound the inner wedge tightly in place from the wider end, creating a strong tension against the glued boards. If the boards to be glued are thin, it will be necessary to place a weight on them to keep them from buckling. When using wedge clamps, remember that the work must be done on a level, clean surface, such as a workbench with a back wall to butt against. Other wedge clamp suggestions are shown in Fig. 3-17.

Tourniquet Clamps. A tourniquet clamp, made from a length of ordinary heavy cotton clothesline, is sometimes more satisfactory for certain clamping jobs than a mechanical clamp, as it distributes the pressure more evenly. Wrap the line around the pieces to be clamped twice and tie the loose ends. Then, insert a short stick (a short dowel or large spike is good) between the line strands and turn it (Fig. 3-18). As you do, pressure is applied to the piece; stop when the desired tension is obtained. Then, put one end of the stick under another part of the piece to hold the pressure. It is necessary to cut the clothesline to the size

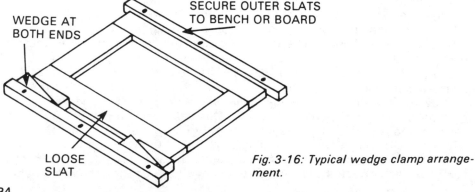

WEDGE AT BOTH ENDS

SECURE OUTER SLATS TO BENCH OR BOARD

LOOSE SLAT

Fig. 3-16: Typical wedge clamp arrangement.

required. When two tourniquets are needed on a single job, each end of the line may be used.

Use and Care of Clamps. When using clamps, first consider your clamping requirements, and then select the clamp best suited to your needs by ascertaining (1) the opening required; (2) the depth required; (3) the strength and weight required; (4) whether or not a full-length screw is essential (if not, the constant hindrance of the screw extending beyond the frame can be eliminated by selecting a clamp with a screw length proportionate to your needs); (5) the type of handle best suited to your needs; and (6) the balance of clamp operating time versus clamping needs, i.e., do you require a spring clamp, C- clamp, bar clamp, or some other type. This appraisal of your clamping needs will assure you of getting the most from your clamps by saving time and money and enhancing the life of the clamps.

With all clamps (except possibly the handscrew type), it is well to remember that heavy pressures crush softwoods. To keep from ruining the workpiece, large blocks are used to distribute the pressure over a wider surface. Provide yourself with a stock of blocks of assorted sizes, thicknesses, and lengths, preferably of hardwood such as maple or birch.

Use clamps in pairs on both ends of the work. This will prevent one end from separating while the other is being joined. For large surfaces, additional clamps are needed.

Apply even pressure on all clamps. Tighten as far as possible with the handle. After a few minutes, take a few extra turns on the clamp handle, if possible. However, avoid pressing in the sides of the workpiece. When closing the jaws of a

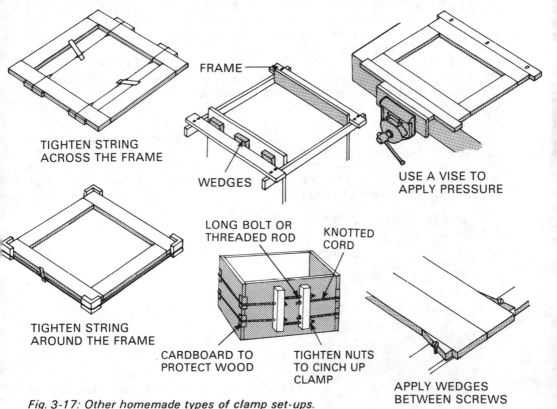

TIGHTEN STRING
ACROSS THE FRAME

FRAME

WEDGES

USE A VISE TO
APPLY PRESSURE

TIGHTEN STRING
AROUND THE FRAME

LONG BOLT OR
THREADED ROD

KNOTTED
CORD

CARDBOARD TO
PROTECT WOOD

TIGHTEN NUTS
TO CINCH UP
CLAMP

APPLY WEDGES
BETWEEN SCREWS

Fig. 3-17: Other homemade types of clamp set-ups.

Fig. 3-18: Tourniquet clamps do a good job of holding.

clamp, avoid getting any portion of your hands or body between the jaws or between one jaw and the work.

Occasionally, lightly lubricate all moving parts for longer service and smoother operation. Make sure there is no oil on any part(s) that will come in contact with the workpiece. The threads of C- clamps must be clean and free from rust. The swivel head must also be clean, smooth, and grit-free. If the swivel head becomes damaged, replace it as follows: Pry open the crimped portion of the head, and remove the head from the ball end of the screw. Replace it with a new head, and recrimp. To maintain wooden screw clamps properly, wipe the wooden surfaces occasionally with linseed oil.

Other Items Needed For Gluing

In addition to clamping devices, you will need a few other items (Fig. 3-19) to prepare and apply the adhesive. In some gluing applications, heating devices are employed to speed up the bonding time or bring about a quicker curing time.

Preparation Tools. When using a water-mixed wood glue, it is very important that the ingredients be carefully measured and thoroughly mixed. Some adhesives give their mixing proportion instructions in terms of weight, others by volume. When it is necessary to measure the weight of an adhesive, use a postal scale for small amounts, a kitchen scale for larger quantities. When measuring by volume, you can employ a cheap metal teaspoon, tablespoon, or kitchen measuring spoon for small amounts; for larger, use a soup ladle or a measured mixing cup or bowl. To mix small quantities of a two-part acrylic adhesive, a glass medicine dropper is handy.

For mixing, use a metal can that is large enough to provide sufficient room for stirring without spillage. Screw-on bottle or jar caps, with their paper or plastic gaskets removed, can be used for very small mixing jobs. For the actual stirring, use a wooden stick or a glass rod. When mixing in a bottle or caps, the wooden stick should be pointed at one end.

Fig. 3-19: A few items that will help to make your gluing job easier.

If you plan to use a flake-type hide glue, an electric temperature-controlled pot is needed. This glue pot is usually water-jacketed like a double boiler for smoother temperature control with minimal risk of overheating.

Application Tools. For many gluing jobs, the adhesive can be applied directly from the tube or squeeze bottle. For some model and delicate woodworking operations, a sharpened, pencil-thin wood stick is better since the point can be used to apply the adhesive to places where the tube tip applicator is too big.

For larger jobs, most wood adhesives can be applied by a roller, trowel (spreader), spray gun, or brush. The latter method is preferred by most homecraftsmen because it is used in the same manner as when painting. The brush need not be an expensive one, and it can be cleaned with water or solvent before the adhesive hardens. Once the glue hardens, it can be a difficult task to clean the brush. Select a width that matches as closely as possible to the width of the coating needed; that is, use a 1" wide brush for a 1" board.

Curing Tools. To speed the setting and curing of some wood glues, a heat lamp can be employed. But when using it, be sure to check the temperature of the wood surface frequently to be sure that it is not so hot that it will catch fire. Should the wood become too hot, the lamp should be moved farther away.

WOOD GLUING PROCEDURES

After the proper adhesive has been selected, the wood surfaces prepared, and the wood joints tested by making a pre-fit trial assembly, you are ready to do the actual gluing procedure. This contains five basic steps: (1) mixing the glue; (2) spreading the glue; (3) assembling the parts; (4) applying proper pressure; and (5) allowing sufficient curing time.

Mixing Glues

When employing ready-to-use adhesives, such as liquid hide, aliphatic, and polyvinyl acetate, no mixing procedure is required. For all powder and two-part adhesives, follow the manufacturer's directions very carefully, making sure to use the proper proportions for mixing. Just mix enough glue for the bonding job at hand so that the batch is fresh.

For all water-mixed wood adhesives (Fig. 3-20), ordinary tap water is sufficient, but it should be at or just a little below room temperature. Chilled water usually tends to lengthen the time needed for the glue to reach working consistency; but, do not use extremely hot water, either. When first mixed, some water-mixed glues appear to need more water, but do not add it . Rather, let the mix stand for 5 to 10 minutes and then resume mixing. If the mix does not turn smooth and workable, add a *little* water; do not overdo it. Before using, make sure all lumps have been completely removed.

When mixing two-part glues, such as resorcinol, the mixing procedure should be carried out quickly. Often, you can stretch the working time a little by placing the container of mixed glue into a bowl of cracked ice (Fig. 3-21). When working with any two-part adhesive, be sure to use twin measuring implements (two spoons, two ladles, etc.) in order to avoid contaminating the material remaining in the two containers. Also, if one component is a powder while the other is a liquid, be careful not to let the powder air-drift into the liquid; keep the containers well separated.

Spreading the Glue

As mentioned in Chapter 2, make a trial assembly to be certain that the joints are well fitted. Do this by clamping all the pieces together. While checking the parts, carefully inspect each one to make sure that all sandpapering has been completed. This will open the pores of the wood, giving the adhesive more hold-

Fig. 3-20: An ice cream stick makes a good mixing device for water-mixed adhesives. It is also good for spreading wood adhesive.

Fig. 3-21: The working time of two-part adhesives can be extended by keeping the mixed glue cool.

ing power. Also, the joints should be square; the more the surfaces touch, the better the grip will be. After all the parts line up square and true, mark all companion pieces with a soft pencil or carpenter's crayon as 1 and 1, 2 and 2, etc. (Fig. 3-22); then indicate front right, front left, etc. (In edge gluing several pieces together, it is a good idea to draw a large X across the face of the pieces.) Once this is done, the different pieces will fit together as they should for the actual gluing without further adjustment; this is important since the assembly time should go as fast and smoothly as possible. The clamps should also be set to the correct opening beforehand to facilitate the gluing operation. Also, have the woodpiece parts and the glue at room temperature, since cold glue and cold wood will not bond properly. No type of wood glue will function properly if it or the piece to be glued is too cold. For best results, gluing and drying of glue should be done in room temperatures of 70 to 75° F.

The common methods of applying wood adhesives are with a squeeze bottle, wood stick or paddle, paintbrush, glue roller, or spray gun. The glue may be applied either to one surface (single spreading) or to both surfaces (double spreading). When applying glue to porous surfaces such as end grain, it is wise to spread a thin coat of glue on both surfaces to be joined; let it stand until tacky. This first coat will dry partly by evaporation, partly by being drawn into the pores of the material. Then spread on a second coat.

Glue from a squeeze bottle should be applied in a zig-zag line. Then, the two parts should be pressed together and moved back and forth to produce an even spread. As a rule, the proper amount of applied adhesive will cause *tiny* beads of glue to appear along the glue line at regular intervals of 2" to 4" when clamped.

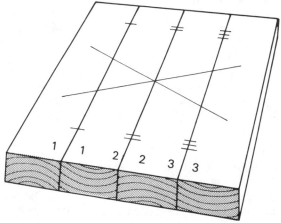
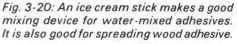

Fig. 3-22: Methods of marking pieces before gluing up.

A starved glue joint will generally result from using too little adhesive or too great a clamping pressure. Too much adhesive will cause squeeze-out and running of the glue down the wood surface. This can be a messy problem, especially when the project is to be finished with stain or varnish.

Assembly Time

Assembly time refers to the time lapse between glue spreading and application of pressure. If you have taken care during the trial assembly period and all parts fit perfectly, assembly of the workpiece should move smoothly. But, there are several factors that will determine the amount of assembly time available. The important ones are as follows:

1. **Type of Glue.** The "set" time of adhesives can vary greatly. For instance, liquid hide glues, which dry or set fairly slowly, have a longer assembly time than aliphatics or polyvinyl acetates. Resorcinols or acrylics have a rather short assembly time.

2. **Glue Formulation.** Frequently, glues with lower solid contents have an increased assembly time.

3. **Glue Spread.** A heavier glue spread increases assembly time.

4. **Temperature.** A lower temperature increases the assembly time.

5. **Substrate.** Use of high pressure laminate or particleboard gives a longer closed assembly time than wood veneers. Ring porous woods, such as oak, ash, or walnut, give longer assembly time than maple or poplar.

6. **Moisture Content.** Drier woods or substrates decrease assembly times.

If glue squeeze-out occurs on the application of pressure, the maximum assembly time has not been exceeded. As long as the glue is wet enough to transfer uniformly to the opposite face when pressure is applied, good strength will result.

Applying Proper Pressure

Most good wood gluing requires the use of some type of clamps. The purpose of the clamps is to bring the members being glued in close enough contact to produce a thin uniform glue line and to hold them in this position until the glue has developed enough strength to hold the assembly together. If the members of a glued construction were to fit together perfectly so that a thin even glue line could be produced, no clamp pressure would be required. But, from a practical standpoint, since machining of stock is never perfect, a certain amount of clamping pressure must be used.

To insure close contact in edge gluing and lumber laminating operations, 100 to 150 pounds per square inch (psi) is usually required. This pressure can usually be obtained by tightening the clamp with your fingers only. Sometimes, it is a good idea to give the clamp a turn or two after a few minutes. But, do not distort the piece; it can be extremely frustrating to break a delicate piece which took a great deal of time to make. In some instances, nails can be used as temporary clamps.

Keep in mind that heavy pressures will crush the softer woods. To prevent injuries of this sort, as mentioned earlier, place small pieces of hardwood, leather, or hardboard between the metal clamping feet and the material being clamped. For very delicate work, you can make blocks of wood with felt or foam rubber strips cemented to one face and use these between the work and the clamps. Also, it is a good idea to put waxed paper between the glue joints and the clamps or clamping blocks so they do not become glued to the stock (Fig. 3-23). If possible, use clamps in pairs on both ends of the work to prevent one end from separating while the other is being joined. For large surfaces, additional clamps are

Fig. 3-23: Using waxed paper or aluminum foil to prevent the clamps or protective blocks from sticking to other wood surfaces.

needed. Apply even pressure on all clamps, but avoid pressing in the sides of the work. If you have a job that requires pressure and the workpiece is too wide for any of the clamps you have available, just open one clamp to approximately its full length and hook its jaw over the jaw of a second clamp. Then, tighten the second clamp to the tension desired, and your problem is solved. Regardless of which clamps are used, there is usually a tendency for long-glued joints to spring apart at the ends. Greater clamping pressure should therefore be applied to the ends than to the center, and the center clamps should be tightened before the end clamps.

After the clamps have been applied, test the job for squareness. Use a damp cloth to wipe excess glue off clamps and stock surfaces. Do this as soon as the material is clamped securely together, so that later sanding and smoothing will be easier and clamps will not get clogged with glue. If you throw sawdust over the squeeze-out as it oozes out of the joint, the sawdust will absorb the moisture of the glue, making it easier to peel or scratch off the excess. The latter can be done by using a sharp chisel along the squeeze-out line. Hold the chisel with the bevel side up. Cutting across the grain (where possible), remove all glue that still remains. Follow this with a thorough sandpapering of all such parts. Give a final sanding to all parts of the furniture with fine and then very fine abrasive paper. Never *wash off* excess glue with water. This would only coat the wood with a thin layer of glue that might show when it dries; it might also make the wood swell, which is very undesirable.

Minor gaps between joints can usually be filled with wood filler after the adhesive hardens. However, if the defect is detected before the adhesive has set, it is a good idea to fill them with a glue-coated wood sliver.

Up to this point, we have concerned ourselves with so-called permanent glue joints. There are two other methods of joining wood: edge-to-edge and face-to-face.

Edge-to-Edge Gluing. There are times when a woodworker needs a board wider than he can purchase in the usual market. When this need arises, the wider board can be made by gluing it up from narrower stock placed edge-to-edge.

The arrangement of the boards for assembly is most important in gluing edge-to-edge. If you place them with all the sap sides at the top, all will cup in the same direction as they shrink and the wide board will be a trough (Fig. 3-24A). Alternate sap with heart sides produce a wavy board which will approximate flatness if warping has been slight or the pieces are narrow; here a little planing will take out the curl (Fig. 3-24B). If, in addition, you arrange the boards with the grain running in the same direction, planing will be easy when the work is removed from the clamps. If the lumber, when assembled, will be stiffened with supporting cleats, as in a table top, boards 6" to 10" wide are suitable, and in natural or

40

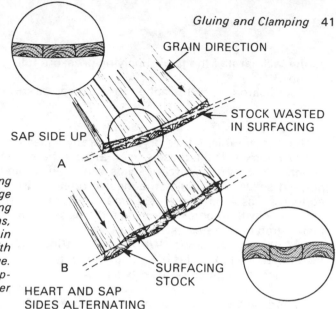

GRAIN DIRECTION

STOCK WASTED
IN SURFACING

SAP SIDE UP

A

Fig. 3-24: Method of laying up boards for edge-to-edge gluing. When edge gluing boards of different widths, place the narrower boards in the center of the panel, with the wider boards at the edge. It is important that the clamping pressure be uniform over the entire glued area.

SURFACING
STOCK

B

HEART AND SAP
SIDES ALTERNATING

stained finishes may even look better. Where extreme flatness is required, as in a drawing board, use narrow pieces. Since the piece must be surfaced, save time by planing all boards to flatness and the same thickness.

Cut pieces 1/2" longer than finished length, arranging them on sawhorses and marking them as described earlier in this chapter. After the pieces are test fitted, spread the adhesive on the edges being bonded. Then, apply the clamps (usually bar or pipe types) alternately, one from one side and one from the other, about every 10" to 15". A slight amount of glue should be visible along the glue line when the correct amount of pressure is applied. This squeeze-out can be wiped off with a damp cloth, or it can be removed by any one of the other methods described earlier. If a board slips out of place while clamping, put a piece of scrap over the joint and knock it into position with a mallet. Carefully check when tightening the clamps to be sure that all corners of the workpiece ride down flat and secure against the clamps, so the surface will be flat with no warping or twisting. If the stock to be glued up is somewhat warped, it can frequently be straightened by clamping all four corners tightly against the bar of the bar clamp, as shown in Fig. 3-25.

Face-to-Face Gluing. Wood built up in thickness by face-to-face gluing is likely to be more reliable than solid blocks; and by facing sap side to sap side, or the reverse, the tendency of one board to warp is offset by the opposite pull of the other board. Wide boards glued in the way described above can be joined face-to-face to increase thickness. Match up boards of about the same grain characteristics; gluing a board having wide annular rings to one with narrow, or a slash board with one quarter-sawed, invites warping as soon as moisture content changes.

As pressure must be applied uniformly, clamps working at the center are necessary. Hand screws are standard for this work, and the maple jaws are adaptable to many conditions of use. Open or close the clamp by holding the handles tightly and rotating it like a crank. For first clamping, set the jaws slightly wide, draw them together with the nearest screw, and then tighten the rear screw. This applies pressure at the point, squeezing surplus glue out from the center toward the edges of the assembly, thus driving out air bubbles. Then, loosen one of the clamps, tighten it toward the back, and apply to the work. This will place pressure on the edges, closing them tightly. Further turning of the rear screw will parallel the jaws, applying pressure full-length. Do this with all the clamps.

Glued boards, when the adhesive is dry, can be surfaced and squared like solid wood.

Curing Time

The rate at which glues dry is generally based on these three factors:

1. **Adhesive.** The adhesive itself affects the speed of set in many ways. An adhesive with a high percent solids will often set faster than one with a lower percent solids. An emulsion adhesive releases its water more easily than an adhesive dissolved in water. Some emulsion adhesives "break" or coalesce more readily than others, causing a faster set. Some with a wet tack will give a faster grab than a non-tacky adhesive.

2. **Adhered Materials.** With a fast-setting glue, joint strength increases faster in the initial stages of setting than in the later stages. With normal gluing conditions, cold glues set over twice as fast on hard maple as they do on ring-porous woods, such as walnut, oak, and ash. Many of the less dense woods, such as pine and poplar, in reality set more slowly than maple, but their lower strength and resultant lower rigidity reduce the stress placed upon the glued joint when unclamped; the required clamp time is, therefore, not as long as for denser ring-porous woods. Most particleboards set slowly and require a long clamp time.

As speed of set is directly related to the drying of the glue in the joint, it can readily be seen that drier wood, with its fast water absorbency, will set faster than higher moisture content wood. Higher wood moisture content will significantly increase the clamp time. The same concept carries over into the effect of exposure conditions on the gluing surface immediately prior to gluing. Though no change might be noticed on a moisture meter reading, the exposure of the surface-to-be-glued to high humidity will slow down the speed of set because these surfaces collect a thin film of water. This is particularly noticeable during humid summer weather. The effect is magnified when machining is done some days prior to gluing.

An ideal glue joint should never measure over several thousandths inch thick. With a poor fit, where this might be increased to .030", as with a rub joint or sloppy dowel, this inclusion in the glue joint of many times the optimum amount of glue will slow the speed of set. As indicated previously, the first water removed from the glue film by the dry wood leaves rapidly. After the wood adjacent to the glue becomes saturated, the water leaves much more slowly, causing the glue to set more slowly. If the dried glue film is visible when the joint is ruptured, the glue line was too heavy to get maximum speed of set.

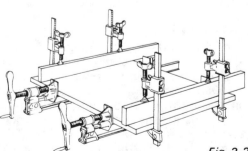

Fig. 3-25: Method of clamping warped boards.

3. **Conditions of Gluing.** The temperature of the wood, glue, and room are important in determining the speed of set. Clamp times at summer temperatures can be one-half that encountered in cold workshops in winter. That is, the higher the temperature, the faster the cure. In commercial woodworking establishments, the most common methods of applying heat are by contact heat with a hot platen and by a high-frequency electrical field. In the home shop, the most common means of using heat is with a heat lamp; but, when using it, caution must be taken not to overheat the joint—this could cause the adhesive to fail.

The use of air circulation may aid in the removal of water from the glue film and speed the set.

GLUE ASSEMBLING OF FURNITURE

Before assembling any furniture piece, you must decide whether it should be assembled all at once or whether it should be glued up in subassemblies. For instance, it is usually better to use the subassembly plan when gluing up chairs consisting of rails and legs. The rails and legs on either end are glued up first, then the entire furniture piece is assembled.

To illustrate how a furniture piece is assembled with glue, take a look at Fig. 3-26, which shows two popular tables with the butcherblock effect. You can alter the dimensions to fit your particular needs, but the basic 18" cube or 15" cube variation will function as an end table, bedside table, or chess and checker table. And the longer table will do well as a cocktail table. The only lumber required for these tables is 2" by 2", graded standard or better. This is inexpensive wood, so it is no financial hardship to buy 25% more than you need to compensate for flaws. Before you begin any construction, inspect all lumber and choose the best boards—those with the most attractive grain and fewest imperfections.

Final finish is a matter of personal choice. Tables can be stained to any shade desired or left natural. They can be varnished, plasticized with polyurethane, or oiled and waxed. If you like the rich, satin luster of an oiled and waxed finish, use a dark walnut stain and Danish oil and paste wax for a hand-rubbed finish.

LAMINATING

Laminating is usually defined as the process of gluing together a number of relatively thin layers of wood to make a heavier piece. In the book, however, we have confined ourselves to the making of decorative lamination (Fig. 3-27A). The same basic technique, however, can be used in the construction of wooden airplane propellors, floor decks, and large structural building beams (Fig. 3-27B). In many of these applications the wood layers are assembled so the grain direction of each layer is parallel to that of the other for greatest possible strength.

To do most laminating work, some type of press is needed. Figure 3-28 shows a jig with a 13" by 13" capacity, which will handle all the laminations in Fig. 3-27A. In use, the jigs should be screwed to a very flat surface so the finished lamination will be flat.

Wax paper laid on the edges of the upper and lower guides will prevent glue from sticking to them. Plastic electrical tape applied to the edges is better yet, because it seldom needs replacement and cannot wrinkle up between the pieces being glued.

Material used for the jigs should be thoroughly seasoned, fairly straight grained, and reasonably flat. Hard maple or birch is preferred, but poplar or pine is acceptable. The step-by-step procedure for making these jigs is as follows:

1. Saw and joint (or hand-plane) boards to size. Round the edges slightly. If tape is to be used, apply it to the better edge of each guide (upper and lower). Two

A

B

C

D

E

F

Fig. 3-26: Butcherblock type tables are easy to build at only a small cost to the home craftsman. While these instructions deal with the cocktail table in detail, they can be applied to cube type tables as well. Each cocktail table requires eleven 8' 2 by 2s of standard grade or better. (Cube tables of 15" or 18" size both require six 2 by 2s.) Cut twenty 36" long pieces for the top and base, eight 15-3/4" pieces for side sections, and six 15" legs. (For each 18" cube table, cut twenty-four 18" pieces for the top, base, and four legs, and four 15" side sections. The 15" cubes require twenty-four 15" top, base, and leg pieces, along with four 12" sides.) (A) Select the ten best 36" pieces for the table top and pre-nail one side of nine with six 8d finishing nails in each. Stagger nails in alternating boards. (B) Apply an even coat of aliphatic adhesive to one side of the unnailed piece and to the side opposite the nails on the first pre-nailed 2 by 2. Join the two together and drive in the two end nails, making sure the pieces are aligned properly. (C) Drive in the remaining four nails and countersink them with a nailset. Laminate the remaining 2 by 2s in this fashion, wiping away excess glue with a damp cloth. Assemble the base in the same manner. (D) Glue two 15-3/4" side pieces to each side of the top and base, leaving 1-1/2"

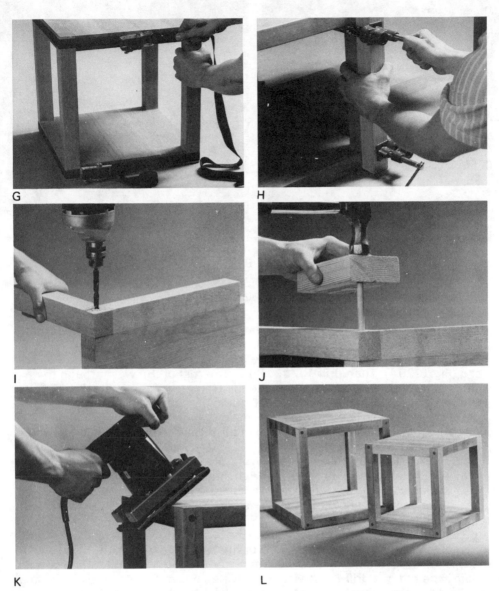

leg notches at the corners and between the two section pieces. Fasten them in place with pipe clamps, and before tightening, check the corners for a flush fit with a scrap piece of 2 by 2. (Cube tables have only one section piece to a side and no center notch or leg, but assembly is basically the same.) (E) After allowing the glue to dry overnight, sand the top, base, and table legs to a smooth, even finish, using coarse, then medium, and finally fine-grit abrasive paper. (F) Apply aliphatic adhesive to the corner notches of the base and top and to the last 1-1/2" of the leg ends. (G) Secure the base and top in place with web strap clamp, and adjust the legs to a flush fit before letting dry overnight. (H) Glue the center legs into place and clamp them fast with pipe clamps. (I) Drill a 3" deep 3/8" diameter hole through each leg and into the side of the table top and base. (J) Insert a 3-1/2" piece of dowel for added leg support. Use a piece of scrap wood as a pounding block to avoid damaging the dowel. Trim it flush to the leg surface. (K) Round all edges and corners with medium-grit abrasive paper. (L) Use finer grade papers on all other surfaces, and as a last step before applying stains or other finishes, polish the table with extra-fine steel wool.

A   B

Fig. 3-27: Various decorative lamination can be achieved in the home workshop (left), but structural types (right) should be left to the professionals.

strips of 3/4" tape should be used on each edge, overlapped about 1/8" at the center.

2. Drill 1/2" holes 5/16" deep in the top frames at the pressure-screw locations (Fig. 3-29A). Then drill 11/32" holes the rest of the way through the top frames at the same locations. Using the top frame as a guide (Fig. 3-29B), drill the shallow 11/32" holes in the upper guide.

3. Embed the 5/16" nuts in the top frame by simply pulling them into the 1/2" holes with a 5/16" bolt or with a piece of all-thread rod and a nut, as shown in Fig. 3-29C.

4. Screw and glue the base to the lower guide. Then, cut the cardboard spacers to size. A tablet back is about the right thickness.

5. Assemble the jig with the clamps (Fig. 3-29D). Drill the 9/32" and 3/8" holes at the location shown in the plans (Fig. 3-28). Assemble the frame with carriage bolts.

6. Cut the 5/16"-18 all-thread rods to the length shown in the plans. Grind or file one end to a shallow point. Grind or file two flat sides on the other end to make it easier to drill the cross-hole. Centerpunch and drill the 1/8" cross-hole, holding the rod in a drill-press vise. Remove all burrs. Then, cut the 1/8" steel rods 2" long. Deburr the ends. Make a small dent in the center of the rod with a hammer. Tap it into the cross-hole in the pressure screw. If it does not fit tightly, make a bigger dent in it and try again. Screw the pressure screws into the jigs.

Striped Laminations

In striped laminations (Fig. 3-27A), the key to an attractive finished product is variety in color and thickness of the woods used. (The thickness of the board you start with becomes the width of the stripe in the finished lamination. That is, all strips are ripped to the same width from boards of different thicknesses and then turned up on edge to be glued together.) Ready-mixed adhesives, such as aliphatics and polyvinyl acetate, work for most laminations.

The straight stripe pattern is made in the following manner:

1. Select a variety of boards for contrasting thickness and color. Sides must be clean and smooth to produce strong glue joints. Moderate warpage will be

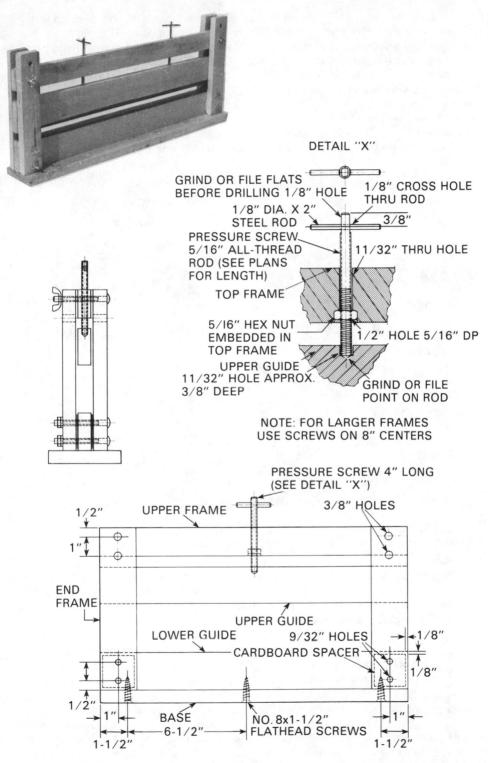

DETAIL "X"

GRIND OR FILE FLATS
BEFORE DRILLING 1/8" HOLE

1/8" CROSS HOLE
THRU ROD

1/8" DIA. X 2"
STEEL ROD

3/8"

PRESSURE SCREW
5/16" ALL-THREAD
ROD (SEE PLANS
FOR LENGTH)

11/32" THRU HOLE

TOP FRAME

5/16" HEX NUT
EMBEDDED IN
TOP FRAME

1/2" HOLE 5/16" DP

UPPER GUIDE
11/32" HOLE APPROX.
3/8" DEEP

GRIND OR FILE
POINT ON ROD

NOTE: FOR LARGER FRAMES
USE SCREWS ON 8" CENTERS

PRESSURE SCREW 4" LONG
(SEE DETAIL "X")

3/8" HOLES

1/2"

UPPER FRAME

1"

END
FRAME

UPPER GUIDE

LOWER GUIDE

9/32" HOLES

1/8"

CARDBOARD SPACER

1/8"

1/2"

1"

BASE

NO. 8x1-1/2"
FLATHEAD SCREWS

1"

6-1/2"

1-1/2"

1-1/2"

Fig. 3-28: Plans for making a simple laminating press.

corrected in the laminating process. Crosscut the boards to the desired length, and joint or saw one edge straight. Then rip the boards to 1" strips.

2. Veneers are best cut with a knife. They may be stacked two or three high to save time. Make light, repeated cuts to avoid splitting them.

3. Arrange the strips in an attractive pattern of the desired width. The strips used for the edges should be at least 1/2" thick so they will not bulge between clamps. Screw or clamp the laminating jigs to a flat surface, after positioning them so they will not interfere with the desired clamp locations. If possible, tilt the entire surface, including the jigs, to a near-vertical position. This will keep glue from running out of the joints before the clamps can be applied.

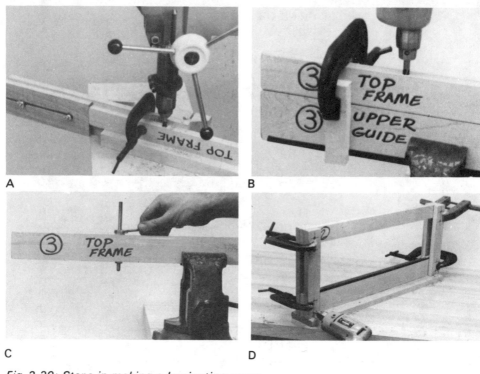

A   B

C   D

Fig. 3-29: Steps in making a laminating press.

4. Apply a heavy spread of glue to one side of the first strip and lay it in the jigs (Fig. 3-30A). Lay each succeeding strip in the jigs after applying glue to one side.

5. Reassemble the jigs and screw them down gently against the work (Fig. 3-30B). Apply clamps (with wax paper between the clamps and the work) and tighten them lightly. Now tighten the jigs down moderately hard against the work and tighten the clamps fully. Glue should squeeze out of all the joints. If not, do not try to take the piece apart now, but use more glue next time.

6. Allow the glue to dry and remove the work from the clamps and jigs. Lay the lamination on a flat surface and build a scrap-wood sanding frame around it. The frame holds the work in position while you are sanding it and keeps the edges from getting rounded off. It should be flush with or a little below the surface of the work. The easy way to do this is to make the frame too high and shim the work up to position with small scraps of cardboard under the corners.

7. Using a portable disk sander or belt sander, remove glue and level both

A   B

Fig. 3-30: (Left) Building up a lamination; (right) jigs and clamps tightened. Better too much glue than too little.

sides of the work with coarse, open-coat abrasive. A disk sander is 5 to 10 times as fast as a belt sander and leaves a flatter surface; it should be used with a flexible rubber or phenolic backup disk and tilted as little as possible. Sand the frame as if it were part of the workpiece. Then, trim the ends square.

VENEERING

The practice of veneering wood surfaces dates back to the ancient Egyptians. In France, during the 1700's, the gifted craftsmen, Hepplewhite and Sheraton, elevated veneering to a true art. Today, home craftsmen can purchase plywood surfaced on one or both sides with a broad selection of veneer woods, thereby eliminating much of the time-consuming gluing and pressing necessary to do an authentic veneering job. Step-by-step veneering, however, still gives the home craftsman a method of producing beautifully finished products with foundations of low-priced wood.

Specialty wood suppliers and craft shops can supply you with numerous veneers of domestic and foreign woods. The usual thickness of such veneers is about 1/28", and they are available in several different pattern arrangements, the most popular being the diamond and herringbone styles. When working with the diamond or reverse diamond pattern, it is essential to cut the four pieces so they have matching grains. All four sections can be cut from one piece of veneer with a straight parallel grain by following the method shown in Fig. 3-31A. If the veneer is too narrow to accommodate the full length of all four sections, the top (or bottom) triangular corners may extend beyond the top (or bottom) edge of the veneer. Complete these sections later using properly-matched triangular patches cut from the waste. Herringbone patterns are cut in the same way and reassembled with alternate pieces displaying opposing grain angles, as shown in Fig. 3-31B.

Cutting thin veneers can easily be done with a sharp knife, heavy scissors, or a veneer saw. Veneer presses, such as those shown in Fig. 3-32, are useful aids in veneering jobs, and constructing one is a simple task. When using the press or

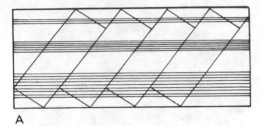
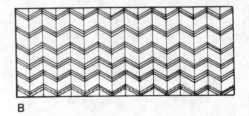

A   B

Fig. 3-31: Common kinds of matching.

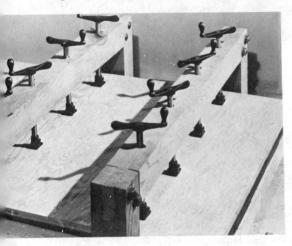

Fig. 3-32: Two simple veneer presses.

working with any sort of pressure screws or clamps, lay the veneer on a flat "caul" or platen and cover it with a second caul. Many home craftsmen use platens of 3/4" plywood in their workshops. The lower platen rests on and is supported by the solid stock bearers or lower crossbars of the veneer press. The bearers upon which pressure is exerted are located directly above these bottom bearers. The centers of the top bearers have a special crowned or arched construction (Fig. 3-33). This guarantees the pressure will be exerted on the center of the glued area first, pushing the glue out toward the edges. The result is a smoother, more uniform bond. The same principle applies when you are using pressure screws, hand clamps, bench clamps, or C-clamps, so be sure to screw down the clamps or screws controlling the center bearer first. It is a good idea to place pads of folded newspaper between the veneer and its caul to catch any squeezed-out glue. Remember that when veneering, application of too much glue will promote "bleed through." Bleed through will make it difficult to properly finish a workpiece, as well as making it difficult to separate the clamping platens or adjacent panels from the veneer face.

Diamond patterns with inlay bands and crossbands or borders around the edges present a slightly more involved job. You may find it easiest to lay out the design on a full-sized paper pattern. Secure the various pieces snugly in place with pushpins or daubs of glue while applying gummed paper to the joints. If you prefer, you can fit the inlay band and crossband borders in place as the work progresses. To use this method, cut the main veneer scant so that after it has been glued in place, waste pieces can be cut away to make space for the inlay bands and borders. To make these cuts, press a cutting gauge against the edge

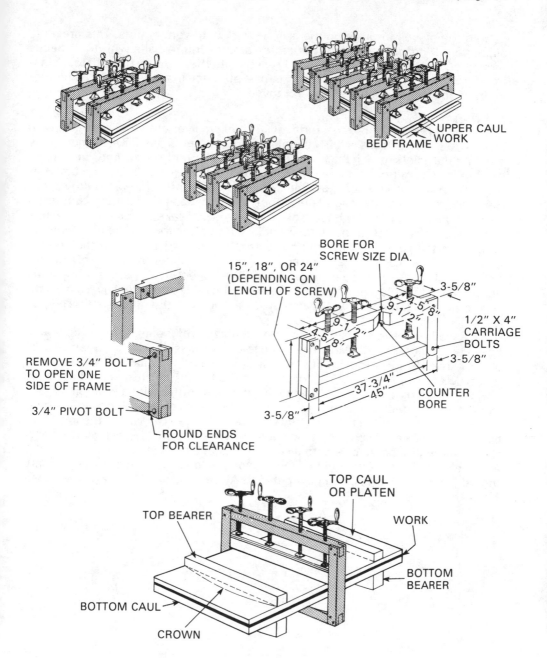

Fig. 3-33: Making and setting up a veneer press.

of the core and simply peel off the waste pieces. As the job proceeds, the corners of bands and borders must be mitered in the same way that the overlap was cut for joining the two veneers in the hammer method of veneering. Take care in matching the inlay band at its mortised corners, being sure the miter is cut through a single portion of wood so that it matches on both sides of the corner. Also, select the crossbanding so that the grains of adjacent strips match at the corners as in a reverse diamond pattern.

Hold the pieces in place on the groundwork with veneer pins. This prevents slippage during pressing. Light brads or picture frame nails with their heads nipped off can also be of some help. Pressed into the cauls by hand before placing the work in the veneer press, they can be punched in after the glue completely dries and the cauls have been removed.

Inlaying

Setting rare woods into the surfaces of solid wood or other veneers is known as inlaying. The process is fairly simple when a router is used. While most inlay banding materials have a thickness of 1/20", their widths vary, and you will not be able to determine the groove width needed until after you choose your materials. To make the cleanest possible cut, use a lefthand spiral bit in the router. Set the cutting depth to slightly less than the thickness of the inlay, and use a straight guide to control the location of the groove. Make a single pass to cut a groove equal in width to the inlay banding, and after the groove is machined, square up the rounded corners with a small chisel. After cutting the groove, miter the corners of the inlay strip and assemble the pieces to be sure of a proper fit before gluing. If all appears to be in order, remove the inlay and apply a small quantity of white glue to the groove. Reinsert the inlay and, after covering it with waxed paper, clamp a strip of wood over the top until the glue has thoroughly dried.

Rare wood block inlays can be mounted in the tops of tables, buffets, decorator boxes, and similar items. Most are made to a thickness of 1/20" and can be purchased in a wide assortment of contrasting or blending colors. When applying this type of inlay, mark out the location on the mounting surface and then cut around the design with a sharp knife. Next, use a router with a straight bit. Set the cutter to a depth slightly less than the thickness of the inlay, and, using a guide, remove as much material as possible from the area where the inlay will be mounted. Remove the remainder of the wood from this area by freehand routing or with a small chisel. Inlays are manufactured with paper on one side. After test-fitting, spread a thin even layer of glue on the unpapered side, insert the inlay into the recess, and clamp it fast. After the glue is dry, sand the paper off of the inlay.

Furniture Repairs and Gluing 4

Adhesives play a vital part in the repair of most furniture pieces. In fact, most repairs just involve using glue, since furniture generally comes apart at the joints. Of course, glue can also be employed to mend splits, relaminate delaminated veneer, patch surfaces, strengthen legs, and revamp binding drawers, among other repairs. Most of these repairs can best be done by employing a ready-to-use adhesive such as liquid hide glue, aliphatic adhesive, or polyvinyl acetate.

Basic Furniture Repair Techniques

For best results with any furniture regluing job, keep the following basic tips in mind:

1. It is difficult, if not impossible to reglue dirty joints or those filled with old glue. Therefore, dismantle the workpiece and clean it. Carefully try to pull apart the loose joints by hand. On tighter joints, use a hammer or mallet, employing a wooden block or thick, folded newspaper to protect the furniture. Joints in very good condition should not be touched.

2. All old paint, wax, dust, oil, grease, glue, etc. must be scraped away or otherwise removed from all surfaces to be glued. Warm vinegar will generally soften most stubborn glue (Fig. 4-1), but it is important to allow the wood to dry before continuing. Be careful not to remove any wood from the joints.

3. The end grain of the joint is one spot where all the glue need not be removed, since most joints are commonly built with clearance between the end of the dowel, or round, and the bottom of the hole. This insures a tight fit at the shoulder. Simply remove the thickest lumps of glue from the end grain with a knife or other sharp tool.

Fig. 4-1: Warm vinegar will usually soften most old glues.

Fig. 4-2: Use a mallet to dismantle furniture pieces to prevent finish damage.

4. Roughen or slash the surfaces to be glued to form a "tooth" for more holding power. Plane, sand, or scrape uneven surfaces to form perfect, well-fitting contact surfaces.

5. Since it is essential to return worn parts to their original places, dismantle (Fig. 4-2), clean, and replace one piece at a time. If you have to dismantle the entire piece, mark the ends of each part and the holes from which they were removed to insure accurate reassembly.

6. Dipping the parts to be glued in warm water and letting them dry completely will open the wood pores and allow the glue to enter more freely. Warming the parts on top of a radiator or simply in the sunshine are other ways to open wood pores.

7. After the parts are clean and dry, test fit them together before gluing. With a tight fit, you are ready to glue. If the joints are a little loose, follow one of the joint tightening methods outlined later in this chapter.

8. After checking to be sure you are getting the proper parts in the right places, apply glue to both joint surfaces and assemble. Apply pressure with clamps or tourniquets to protect the finish from scratches. Waxed paper under the wood pieces or pads will catch any glue forced out by the pressure. Since glue will not stick to waxed paper, cleanup is easy.

9. After gluing and clamping, wipe the glue which is still soft from the finish. Use a smooth chisel-edged stick to clean around the joints and tight places, and rub the rest of the piece down with a clean, damp cloth. If the piece must be moved to another position before the glue has set, wind a piece of string around the joint several times and knot it to prevent the new glue from running out of the joint. Remove the string before the glue hardens completely. Later, any hardened excess glue can be carefully removed with a knife without damage to the surface.

10. Pieces with many glued joints should be placed on a flat surface and tested for alignment before the glue completely hardens because this is the only time adjustments can be made. The glue must be completely dry before the clamps or tourniquets are removed. Any glue remaining on the surface will produce spots beneath a clear finish, so now is the time to check for and remove it.

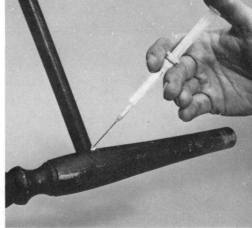

Fig. 4-3: Chair rungs and similar joints can be reglued even when it is not feasible to pull them apart. Drill a small hole into the joint and inject the glue with an oil can (left) or glue injector (right).

While it is a good idea to dismantle a furniture piece before regluing, some really old antiques, especially some rung-type chairs, and furniture held together by wooden pins or wedges should never be completely taken apart. There is a good reason for this. Before glue was commonly used in furniture construction, chair rungs were fashioned out of dry wood with a bulb on each end. These were tightly fitted into the holes of green, unseasoned legs. As the green legs seasoned and shrank, they formed a tight joint with the rungs. With age and wear these joints can become loose, and now they cannot be dismantled without damage.

Loose joints of this type, however, can be reglued in several ways without dismantling them. One good method is to work the glue well into the loose joint using a toothpick. Try to position the piece so that the glue can flow freely down into the joint.

By drilling a 1/16″ hole at an angle to or alongside loose joints, glue can be forced into them with a small oil can (Fig. 4-3), plastic squeeze bottle, or a special tool known as a *glue injector*. (A glue injector is actually a syringe with a hollow needle; squeeze the plunger and the glue comes out.) Inject glue into the joint until a squeeze-out appears. Then, clamp it fast, wipe clean, and let dry.

Mending Split or Broken Parts

Parts which are split or cracked, but not broken into individual pieces, should be repaired without being separated if at all possible. A large percentage of the cracks in wood furniture may be glued and clamped, the result being a permanent repair. Dirt, old glue, and paint must be removed from the crack with a narrow-bladed knife, pin, or thin tool before this type of repair can be made. Dragging an old hacksaw blade through a straight crack is a very good way to clean it. Position the teeth down and pointing toward the worker. Blow out the loose material.

Cracks near an edge should be widened by gently driving in several soft wood wedges, one at a time. When the crack is wide enough, insert the glue, remove the wedges, and clamp tight. Cover the area with waxed paper and place flat sticks under the clamp jaws to protect the wood surface. Tourniquets may be used in place of bar clamps. Or, if you prefer, use brads to draw the crack together after the gluing. Carefully use a small hammer to drive the brad into the board

Fig. 4-4: A broken rung can be repaired by filling the break with glue, pressing the parts back together, and then binding it with a tight wrap of masking tape.

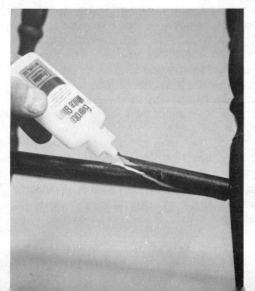

edge, then sink the head below the wood surface with a nailset. Fill the hole with stick shellac, wood putty, or dough the same color as the wood finish, and carefully smooth off the rough spots with a knife or fine abrasive paper.

Cracks farther away from an edge should be thoroughly cleaned and then tested to see if they can be brought together. Use a strong bar clamp with blocks of wood beneath the jaws for surface protection to see if this can be done. If the crack can be drawn together, apply glue and then clamp.

Most broken furniture repairs can be separated into two general categories: supported and unsupported. Broken parts in unsupported repair are simply glued together, and you must rely on the strength of the glue and wood to make a satisfactory mend. For supported repairs, dowels or other types of supporting devices are used to strengthen the repaired parts. Use common sense when deciding whether to glue or use a dowel in the repair. Remember that the more gluing surface there is between the two parts, the stronger the glue joint will be. Small, flat surfaces fastened together by glue make the weakest joints and need structural support. An example of this would be if you tried to glue pieces of 1/2" dowel end to end.

But, if the crack in a split chair rung runs nearly the entire length of the part, a very large surface for gluing exists, and an unsupported repair will usually be successful.

Consider the amount of stress the part will undergo during normal use before deciding on the type of repair. The arms of a dining table chair, for example, take plenty of stress, and dowelling is a good idea whenever possible. But, where stress is less, a simple gluing may be appropriate. For example, broken rungs frequently can be repaired just by filling the break with glue, pressing the parts together, and then clamping (Fig. 4-4).

In all gluing jobs, clamp the repaired part with sufficient pressure applied in the right direction. Allow plenty of time for the glue to dry, being sure to take into account any cavities in the joint which may have made it necessary to apply a thicker coat of glue than normal.

Save all the pieces of a broken part because you will need them when you glue the piece back together. If some of the pieces are missing, fill the void with glue or insert bits of wood or wood putty.

Dowel Pin Repairs. Dowel pins holding furniture pieces secure sometimes snap, leaving one or both ends in the holes. Or, dowels used to secure drop-in leaves to slide-out extension tables often splinter or otherwise become damaged. To make most dowel pin repairs, you must first bore the dowel out of the hole (Fig. 4-5). Use a drill with straight-shank bits slightly smaller than the diameter of the dowel and only drill to the depth of the dowel, since it is quite possible to bore a hole too deep or even drill through to the other side. Since boring through the hardwood dowel is tougher than boring through the softwood of furniture, it is possible to tell by feel when to stop. Force out what remains of the dowel with a small chisel or knife, being careful not to enlarge the hole. Flush the remaining glue with vinegar before selecting a new dowel which fits the hole snugly. Spiral or straight grooved dowels are the best because they allow excess air or glue to escape after the dowels are inserted. If the parts of the joint fail to come close together, the dowel may be too long. If this is the case, cut a piece off one end, round the cut with a knife or abrasive paper, and then follow the directions for gluing.

Mending Split Legs with Caster Inserts. Loose casters can cause furniture legs to split, so it is best to practice a bit of preventive maintenance by tightening any furniture casters you may own with new plastic inserts or with plastic inserts and new caster assemblies.

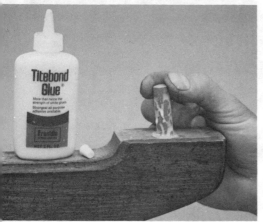

Fig. 4-5: Steps in replacing a dowel: (Left to right-top to bottom) Mark the exact center of the old dowel with a sharp punch. Then, drill out the old dowel, in both pieces if necessary. Squeeze some glue into the dowel hole and insert the new dowel. Apply a thin coat of adhesive around the dowel and fit the pieces together.

To repair splits in legs, force glue between cracks and wrap the leg tightly with masking tape (Fig. 4-6). Rebore the hole into which the caster is inserted with a drill the same size as the hole, to make it clean and smooth. Using a dowel the same size as the hole you have just drilled, coat it with adhesive and slip it into place. After the glue has dried, redrill the caster insert hole, making sure it is the same diameter as the shaft of the caster. Even with the wire clamp in place, an oversized hole may cause the caster to split the leg again. A wobbly leg can cause the caster to split it when weight is applied, so as an added precaution, be sure all legs are firmly secured to the piece of furniture.

Tightening Loose Furniture Joints

Loose joints are most commonly found in chairs, sofas, and tables. This condition generally results from heavy use and movement. In addition to rough usage, joints may also become loose from lack of atmospheric moisture or from shrinkage of the wood. Assuming that a permanent repair is desired, the joint must be tightened properly. Depending upon the prevailing conditions, it may be advisable to reinforce a joint with added mechanical devices. The simplest techniques for tightening round and square or rectangular furniture joints are as follows:

Tightening a Joint with Thread or Cloth. If the piece is not too loose in its socket, it can often be tightened by wrapping thread around the loose part. To repair a loose rung, for instance, wrap the ends with a single layer of thread (Fig. 4-7). Coat the thread with adhesive so that it will properly hold onto the wood. The thread gives the ends of the rungs just enough added thickness to make the

Fig. 4-6: Steps in fixing a loose socket caster.

joint between the rung and the socket tight. When the adhesive on the thread dries, coat the socket holes with glue and insert the rung. Clamp until glue dries.

Another similar method that can be used on either a round or square joint is to cut some cloth strips narrower than the end of the part to be inserted into the hole. Place these cloth strips, in the form of a cross, over the end of the piece (Fig. 4-7). Since the cloth stretches when being inserted into the hole, trim the material on the sides from one-half to three-quarters the depth of the joint. Apply glue and set the joint together. Should the cloth protrude out of the joint, closely trim it with a razor blade and, with a damp cloth, wipe off any excess glue.

Fig. 4-7: A layer of thread (left) and thickness of cloth (right) may be sufficient to tighten a loose rung.

Plugging a Round Hole with a Dowel. The receiving holes often enlarge—for example, sway causes the legs to "work" in the holes—and must be tightened. One way to do this is to remove the loose piece and glue a piece of hardwood dowel of the same size into the receiving hole. When the inserted section is dry, cut the dowel off flush with the surface and drill a new hole for the part to be inserted.

Cross-doweling can also be used to tighten a loose rung. Remove the rung from the chair; clean out the socket hole and the turned end. Apply adhesive to both the rung end and socket hole. Insert the rung and clamp. Then, drill a 1/8" hole through the leg and the turned end of the rung (Fig. 4-8). Insert a small dowel, well glued, into the drilled hole. When the glue has dried, cut the dowel flush with the surface and sand smooth.

Nails have also been used the same way, with the ends filed flush.

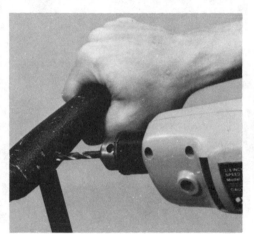 

Fig. 4-8: Cross-doweling is a good method of fastening a loose rung.

Tightening a Joint with a Screw. If the joint is one where the hole does not go all the way through the receiving member, as is the case with stretchers and some mortise-and-tenon joints, it may be tightened and strengthened by inserting a screw through the base of the hole and into the second part. By countersinking the receiving hole from the outside, the screw head will be beneath the surface when driven into the part to be tightened. The countersink can be plugged later with a small piece of dowel, glued in place, smoothed, and finished.

Tightening a Joint with a Wedge. Thin wood wedges (hardwood is best) may be employed to widen a round or square piece so that it will fit snugly against the sides of the hole into which it goes. To accomplish this, cut a saw kerf (slot) in the end of the piece with a backsaw or a dovetail saw (Fig. 4-9). Be careful not to split the wood or let the saw wander off square. The saw kerf should be fairly deep, but should not cut into the exposed portion of the piece.

With a sharp knife, cut a thin wedge as wide as the receiving hole. Apply glue to both the kerf and the receiving hole. Start the wedge into the kerf, set the piece into the receiving hole, and drive it into place with a soft-faced mallet. If the thickness and length of the wedge are correct, it will hit at the base of the holding hole, widen the cut slot, and thus hold the piece tightly in place. The piece must, however, be clamped together (Fig. 4-10) until the glue dries.

When the piece to be widened (chair leg, rung, tenon, etc.) is to fit into a hole that goes all the way through the holding piece, cut a slot, apply glue, and assem-

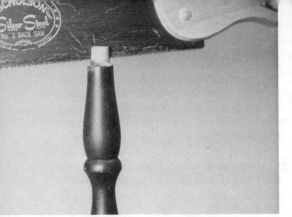

Fig. 4-9: Method of inserting a wedge into a rung.

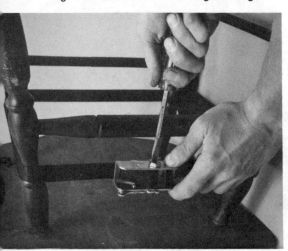

Fig. 4-10: Two methods of clamping chair rungs: (left) strap clamps and (right) tourniquet or rope clamp.

ble the pieces. Then, drive in the wedge, allow the glue to cure, trim off the excess material, and sand the surface to a smooth finish.

Tightening a Square Joint with a Shim. Shims are a proven method of tightening mortise-and-tenon and similar style square-sided joints. Best made of hardwood, these shims may be of even thickness or slightly tapered, but they should be as wide as the receiving hole. Apply the glue to the hole and carefully drive the shim into place.

Adding Corner Blocks. Occasional and dining room chairs, which have cloth-covered, padded seats, must depend almost entirely upon the strength of the joints in the frame of the seat to hold them together. When such chairs are subjected to heavy strain, there is a greater chance of joints coming apart than is the case where chairs are constructed with solid seats. The padded-seat style chair and similar styles of furniture, such as overstuffed chairs and sofas, should have additional support in the form of screws inserted into the frame. Even these screws do not always afford the extra strength required, and additional support is frequently needed.

The best method of repairing and strengthening loosened joints in such furniture is by means of hardwood corner blocks. First, remove the padded seat by taking out the screws on the underside. Then, cut triangular-shaped blocks to fit into the four inside corners of the seat frame. They must fit evenly into the cor-

60

ner, with all edges flush against the frame. Ill-fitting blocks are worse than use-less—adding weight without really strengthening the frame. When possible, the triangular blocks should be at least 1-1/2" thick and the side edges over 3" long. To prevent splitting, drill clearance holes at an angle through the hypotenuse of the triangular block for the screws that are to go into the seat frame, and coun-tersink for the screw heads. Apply glue and screw the blocks into place. Once the glue has dried, replace the padded seat with screws.

If a corner block is missing, remove one of the other blocks and copy it. Then replace both. If a metal brace is missing, it can be replaced with a metal brace of the same design or with a wooden block.

Working with overstuffed furniture is more difficult since the padding, and sometimes the springs, must be removed. If you are not qualified to do this type of work, it is best left to a professional.

Securing Furniture Tops to Frames. Furniture tops usually have to be re-moved to be repaired. Turning the object over will most likely show you how it is fastened. The usual systems employ wooden fasteners with screws, recessed screws in the aprons, glue blocks, and the corner bracket with wing nut. Except for the latter, simply turn out all the screws and the top will usually come free. In the case of the corner brackets, there is a hanger bolt at a 45° angle in the leg it-self, so it slips through the center hole in the corner bracket where it is tightened with a wing nut. If the leg is loose, it can be tightened by giving the wing nut a turn or two. To take off the legs, remove the wing nuts. The apron frequently found fastened to furniture tops can usually be taken off by removing the screws holding it in place.

Split tops can be repaired as previously described on page 55. Warped tops can be repaired by a heat and water treatment (Fig. 4-11). That is, dry out the convex side of the board and/or dampen the concave side. A heat lamp makes a good source of heat, while a wet sponge or towel can be used to apply moisture. Once the board is no longer warped, seal the surfaces with a desirable sealer.

If this method does not solve the warpage problem, make a series of parallel cuts with a power saw, each about one-half the thickness of the top (Fig. 4-12A). These saw kerfs should run with the grain, but should not go to the edges where they would show. Keep the kerfs 3" to 4" apart, and do not make cuts in areas where fasteners are located.

Fig. 4-11: The moisture/heat method for re-moving warps.

After the cuts are made, clamp the top flat, keeping the clamping devices away from the saw cuts. Then, cut wood splines to fill the kerfs. If the top surface is concave (curving upward), cut the splines to fit snugly, since they must act as a filler to keep the board flat. If the top surface is convex (curving downward), make the wood splines a little wider than the kerfs and plane to a slightly wedged cross section. Thus, when the wedge splines are tapped into the kerfs, the pressure they tend to exert will flatten the top.

With either a concave or convex surface, insert the splines with adhesive (Fig. 4-12B), wipe the surface clean, and allow the glue to cure completely before removing the clamps. Sand the splines flush and apply a sealer to keep the moisture content fairly constant.

A   B

Fig. 4-12: (A) Grooves are scored in the underside of a warped top, in areas where they will not show. (B) Then, wood splines are inserted, glued, and planed off smooth.

For securing the repaired boards to the top frames, proceed as follows:

1. If a top consists of two or more boards, scrape any old glue from the edges and roughen them slightly, so that the new glue will have a "tooth" to hold onto.

2. Fill any screw holes with wood putty. Be sure to pack it well into the hole, and then allow it to completely dry.

3. Turn the boards upside down on a flat surface and place them side-by-side in matched grain position over waxed paper. Apply adhesive to the contacting edges and draw the boards together with a wedge clamp. On small tops, clamping is best accomplished with bar clamps, using wood blocks under their jaws.

4. Place the frame upside down on the top boards. Locate its exact former position from marks left at the edge of the frame. Replace all screws that previously held the top to the frame, and tighten securely. In addition, install glue blocks at the joint between the top and frame, at each end of the frame, and along the frame sides. Besides being glued in place, the blocks should also be held with screws. Drill clearance holes through the blocks so that the screws will not bind in them, and then countersink the holes. Use flathead screws to take full advantage of the pull against the head when the screw is tightened.

Repairing Drawers

Since furniture drawers are subject to extra hard usage, their joints frequently become loosened. Should this occur, the drawer should be taken apart and reglued. To accomplish this, turn the drawer upside down and remove the bottom.

It slides into a groove from the back and is often held in place with a small brad driven through the bottom and into the end of the piece. Care must be observed in removing the sides if they have been connected to each other with dovetail joints. To release the strain, tap on a small, flat board held against the drawer side and next to the joint.

After thoroughly cleaning away the old glue, apply adhesive to both members, reassemble, and clamp. Before gluing the bottom in place, insert it into the furniture piece to make sure that the drawer fits properly and slides in and out smoothly. Once this has been determined, you can glue the bottom. Weak drawer corners can be strengthened with corner blocks (Fig. 4-13A).

If the drawer drags or sticks along its sides or its slide, it may have to be sanded or planed. A slight drag may be relieved by rubbing paraffin (Fig. 4-13B) or furniture wax on the sides or slides. If the drawer is too loose vertically, build up the slide by gluing thin strips of wood to it.

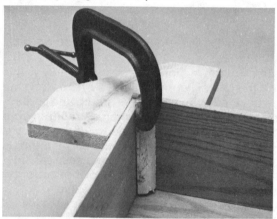

A   B

Fig. 4-13: (A) Brace weak drawer corners with blocks of wood. (B) Rub paraffin on drawer runners to make them slide easily.

Patch Repairing

Mild scratches which are confined to the finish and do not extend into the wood can be easily repaired. For example, a minor abrasion on a varnished surface can usually be removed by brushing turpentine on the damaged area. The turpentine liquefies the varnish, which then flows into the scratch and re-hardens in a short time. The same technique can be employed for lacquered surfaces, except that lacquer thinner is used in place of the turpentine.

Deeper scratches can, as a rule, be repaired with shellac, applying it as directed by its manufacturer. However, when the damaged area extends deeply into the wood, the blemish should be cut out and replaced with a patch of veneer or solid wood whose grain and coloring closely match the original wood.

The shape of the patch is important. To achieve a good grain match, for instance, the lengthwise ends of the patch should be cut at a 45° angle to the grain. If an end edge cut is at right angles, there will be noticeable evidence of the repair. Figure 4-14 illustrates several shapes that will be adequate for most furniture surface patching jobs. You will note that where the end edges of the patch cross the grain, they do so at an angle of 45°, and the side edges run with the grain in long patches. Make a cardboard template of the patch shape and scratch its outline on the surface to be repaired.

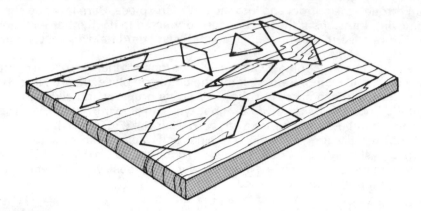

Fig. 4-14: Patch shapes that will accomplish most surface patching jobs.

To dip out the "grave" for the patch, cut down along the scratch marks with a sharp chisel. Then, with a mallet, hit the chisel lightly into the surface along the same lines. Hold the chisel with the beveled edge toward the patch and perpendicular to the surface. Cut square corners. The surface included within your lines is to be cut down to the correct depth. Actually, this depth of the grave is important. When a shallow patch is to be glued in place, its upper surface should be slightly higher than that of the surrounding area. This raised surface can later be sanded down to the original level. Disregard the chisel marks on the bottom of the grave. They actually increase the holding power of the adhesive.

To glue the patch, apply the adhesive to the edges of the patch and the sides of the hole (Fig. 4-15). Press the patch into place, leaving its surface a bit above the surrounding wood. If the grave has been cut too deeply, insert a small brad or pin in the cracks on opposite sides so that the patch will not settle too far into the hole. Before the glue has had a chance to dry, scrape out some of the glue with a pin a bit below the surface. Remove pins and brads used to raise the patch before the glue has hardened. Let the patch set for at least 24 hours in order to make certain that the glue is hard. When you are sure that the glue has set and can take some work, sand the raised patch down to the level of the surrounding wood. Do not press down too hard on your patch as you may damage it. Sand with the grain, using extra fine abrasive paper. Be very careful and deliberate during this phase of the operation. Hasty and careless work here may result in either a completely ruined patch or a patch with clearly defined outlines calling attention to an unsightly repair. The last steps are to fill in the patch outline with a wood filler, which can take a stain, or with a stick shellac of the proper color.

When patching furniture surfaces, your intent is usually to refinish the entire piece. In such cases, you can use different materials for the new finish, especially when you plan to stain the entire surface. But when the patch alone is to be given a finish, it is best to use the same materials that were used for the original finish of the piece.

Repairing Veneer Surfaces. The veneer on older pieces of furniture often comes loose and peels away from the solid core. These curled, wrinkled, and/or blistered segments of veneer are not only unsightly, but, if left untreated, will break off, either leaving a permanently damaged piece or necessitating extensive and time-consuming repairs.

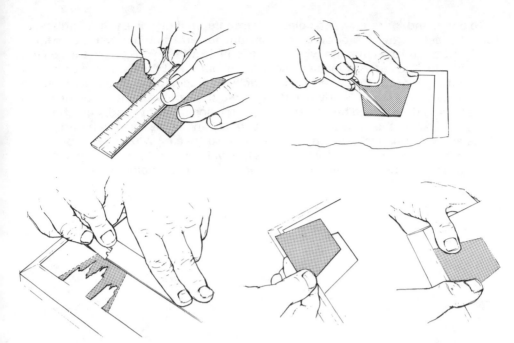

Fig. 4-15: Applying a wood patch.

When repairing loose veneer, the old glue must be scraped out from the base wood with a small, sharp knife or razor blade. Work the cutting tool in under the veneer as far as possible, being careful not to break the surface. If all the glue does not come out, apply a solution of vinegar and hot water under the loose veneer. The solution will eventually melt the glue which then can be scraped away. Wait until the area has dried completely, then reattach the veneer to the base wood by using an aliphatic or polyvinyl acetate glue. Wipe off as much excess glue as possible. Lay plastic sheeting or waxed paper over the repaired area to prevent excess glue from sticking to whatever clamping device you use. After the adhesive sets, carefully wipe away any excessive adhesive with a damp cloth. If it is not possible to use clamps to secure the veneer, masking tape can be placed over the repaired area. If neither clamps nor masking tape can be employed, the patch can be secured by using a weight, such as a sandbag. Allow the adhesive to set for at least 12 hours before removing any securing agent from the repaired area.

Even with the most careful clamping techniques, you may end up with a bubble or two in the veneer. If you have a glue injector, flattening these bubbles will be a simple job. This device, mentioned earlier in this chapter, is simply a hypodermic needle designed for "shooting" glue into small or tight places. Aliphatic and polyvinyl acetate adhesives are usually too thick for use in glue injectors, so mix the glue with an equal amount of warm water or solvent before placing it in the injector. Clean the adhesive out of the injector with vinegar and warm water so that you may use it again.

Plastic Laminate Repair. Rust rings, minor scorches, or black marks from cooking utensils left on plastic laminates can frequently be removed by scouring powder. But, for more serious damage to the surface, patching of some type is generally necessary. If the laminate is a standard color or pattern, a match should be no problem. An edge-to-edge patch is simple to make and is seldom noticeable.

To patch and/or re-cement a plastic laminate, first dissolve the old adhesive (using a lacquer thinner or other appropriate solvent) so that it can be scraped up. Then, pry up the laminate with a small, sharp knife so you can get the solvent underneath. The patch is made by scoring an outline of the damaged area, and then cutting it out with a fine-toothed hacksaw blade. Cut carefully beyond the scored mark, and then file down to the desired size. Draw an outline of the patch pattern onto the surface being repaired and, using a sharp knife, cut a hole for the patch. Apply the adhesive to both surfaces and press the patch into place. Wipe away any excess glue, and clamp the bond, placing waxed paper between the clamp and the patch so that excess glue will not stick to the clamp. If a contact adhesive is used as described in Chapter 5, no clamping is necessary.

Contact Adhesives and How They Can Be Used 5

Contact cements are defined as adhesives which are applied to two surfaces to be bonded together, the adhesive substances being allowed to air dry until they have little residual tack. They are then capable of adhering to themselves instantaneously when the coated surfaces are joined to form a permanent, non-adjustable bond under moderate pressure. They are one of the most popular types of adhesives among do-it-yourselfers, accounting for approximately 28 percent of all consumer sales.

In addition to being employed in the installation of laminated plastics on core stock—their principal use—contact adhesives may be used for bonding hard-board, wood, particleboard, leather, cloth, cardboard, plastic foam, ceramics, steel, aluminum, cove bases, ceiling tile, gypsum wallboard, weatherstripping, rubber and vinyl auto trim, and other similar materials.

TYPES OF CONTACT ADHESIVES

There are three basic types of contact adhesives: solvent-, chlorinated-, or water-based.

Solvent-Based Contact Adhesives

Solvent-based contact adhesives are very fast drying; they dry to a tack-free surface in 5 to 10 minutes. They develop unusually high "green" strength and bond on contact, requiring no clamping or prolonged pressure. Solvent contact adhesives are used primarily to bond high pressure laminates to plywood, hard-board, and particleboard. They are also highly heat and water resistant, and they are available in both spray and brush grades. The latter grade can be brushed, troweled, or roller coated without balling up. But, contact cements of the regular or conventional solvent type are usually composed of a high percentage of solvents (70 to 90% by volume) and a low percentage of solids (10 to 30% by volume). Most of the solvents used in these contact adhesives are *very* volatile. This means that their vapors are *extremely* flammable and, in some instances, toxic.

For this reason, when using solvent contact cements, do not smoke; extinguish all flames and pilot lights; and turn off stoves, heaters, electric motors, and other sources of ignition during their use and until all vapors are gone. Do not use on areas where static electrical sparks may be generated. Keep the container closed when not in use. Children should be kept away from work areas since heavy vapors concentrate at floor level. Prevent build-up of vapors; open windows or doors, and use only with cross ventilation. Avoid breathing vapors; vapors may cause light-headedness. Exposure to fresh air should dispel this symptom. Also avoid contact with the skin, eyes, or clothing; wash thoroughly after handling. In case of ingestion or contact with the eyes, call a physician *immediately*.

Recent rulings by the Consumer Product Safety Commission of the Federal Government banned *extremely* flammable solvent system contact cements from the consumer markets. This was done to help provide a measure of safety to the consumer. Most leading manufacturers have completely removed all extremely flammable solvent contact cements from the market; a few still produce these adhesives for industry or with less volatile solvents.

Chlorinated-Base Contact Adhesives

Chlorinated solvent contact cements are very fast drying, develop excellent green strength, and can be brushed, troweled (fine toothed), rolled, or sprayed. In spite of their pungent chlorinate odor and their cost which is higher than the regular solvent cements, the chlorinated solvent contact adhesives are becoming the most popular with the do-it-yourselfer. In addition, chlorinated contact adhesives are nonflammable, making them particularly important where local fire codes or insurance requirements dictate the use of a nonflammable contact cement.

Chlorinated contact cements will dry to the touch in approximately 5 to 10 minutes. Computing accurate coverage figures for contact cements is most difficult and at best represents only an estimate of what one might expect. Many variables affect these coverage figures, such as porosity of the substrates, method of application, efficiency of application, etc. Best results and optimum performance can be derived from most contact cement formulations if you strive for an approximate 6.00 mil wet film thickness on application. This will yield an approximate 0.75 to 1.00 mil dry film thickness. Film thicknesses of this type will utilize approximately 50 pounds of contact cement per 1,000 square feet of application. Thus, 1 gallon of chlorinated contact will cover approximately 250 to 300 square feet of surface area. Bear in mind, however, that these figures will vary; an extremely porous surface would reduce coverage considerably; a spray application could increase coverage; the operator or applicator could increase or decrease coverage depending upon his efficiency. Therefore, the above should be considered as an average coverage for most applications, and each application should be evaluated for more specific and accurate computations. It is not recommended that these products be thinned.

Solvents such as toluene, xylene, or other clean-up materials recommended by the manufacturer may be used to clean your tools and equipment. For best results, wherever possible, soak the parts in solvent a few minutes if the adhesive has dried. Scrub the parts clean and rinse them in clean solvent. When using a contact cement and clean-up solvents, be sure the work area is adequately ventilated.

Since most water-based contact adhesives contain alkaline hydroxide, keep them out of contact with your eyes. If splashed in an eye, wash it out very well with water for at least 15 minutes. Call a physician *immediately*. Wash your hands thoroughly after using the adhesive. Prior to rewearing, launder clothing which has been contaminated. Keep the cement out of the reach of children.

Water-Base Contact Adhesives

Water-base (sometimes called neoprene latex base) contact cements are nontoxic and nonflammable. They are particularly well suited for use with plastic laminates or gypsum wallboard in areas where the hazard of solvent-based adhesives would be prohibitive. Water-base adhesives are also excellent for use with polystyrene foams and other plastics that would be subject to attack from solvent systems. They may also be used in many other applications where an immediate bond is required or where two relatively nonporous surfaces are to be joined. Do not use them, however, on any metallic surfaces.

Water-base contact adhesives have excellent resistance to heat and water. They can be applied by brush, carpet stipple roller, fine-toothed trowel, or spraying. They usually cover in one coat except for extremely porous materials (cloth, canvas). Water-base cements generally will not mar varnished, lacquered, or painted surfaces. Spills can easily be sponged off with water when still wet. In fact, equipment can be cleaned with water if the cement has not dried. Hot water or steam applied to the dried films will sometimes allow the cement to be rolled from the surface. Toluene or chlorinated solvents will dissolve the dried film. It is not recommended that these products be thinned.

The single largest disadvantage of water-base contact adhesives is that they are slower drying than either regular solvent or chlorinated types. Air drying for up to an hour is often required before contacting the two surfaces. (Color changes will many times indicate dryness.) Actually, with many makes of water-base contact adhesives, working time up to one week before combining can be used. Longer working times usually require higher J-roller pressures. These adhesives give up to twice the coverage of solvent contact cements. One gallon spread at the rate of 4 mils will cover up to 400 square feet of laminate and substrate; it will install up to 200 square feet of plastic laminate. Remember that these products are not freeze/thaw stable. Temperatures below 40°F will break the emulsion, making it unusable. Contact with acids (even a weak acid) can irreversibly precipitate the resin and render the product useless.

There are some acrylic contact cements available to commercial or industrial users. Because of the high pressure required to bond this type of contact cement, they are generally not yet available to the do-it-yourselfer.

LAMINATING WITH CONTACT ADHESIVES

As stated earlier in this chapter, bonding laminated plastics to counter surfaces and furniture pieces is the No. 1 use of contact adhesives. Regardless of the type of contact adhesive, the basic information contained here holds true.

Laminated plastics are available in a variety of decorator colors, attractive motifs, and various patterns. New methods of texturing even make it possible to get laminated plastics with the feel of leather, wood, or slate. Some plastics have a matte finish that will not show scratches, but it is a little harder to clean than the shinier finishes. The standard countertop grade is 1/16" thick, wears extremely well, and is not injured by grease spatters. However, it should not be used as a cutting surface, and hot utensils or small appliances should not be set directly upon it.

Most laminated plastic sheets come in widths of 24", 30", 36", 48", and 60"; lengths are 72", 84", 96", 120", and 144". Any combination of width and length may be ordered; actual sizes are usually slightly oversize so you can obtain a desired rectangle by trimming after bonding. When figuring the amount of material needed for a countertop, remember to avoid joints as much as possible by using the longest lengths available. For example, if you are going to build a 9-1/2' long top, buy a 10' length of base core and laminate rather than joining an 8' piece with a 1-1/2' piece. If the countertop has to turn a corner, you will probably have to join pieces.

Most laminating work is done with plywood or particleboard core. Plywood is relatively easy to work with, is extra-strong, and provides an excellent surface for bonding decorative laminates. When strength is a concern, pick plywood (preferably 3/4" A-D grade; A is the better side). If a part is nonstructural, consider using less costly particleboard (which also is an excellent material for core stock).

For hand cutting plywood or particleboard, use a fine tooth crosscut saw and cut plywood with its good side up. For cutting plywood with a *sabre saw* use a plywood blade and cut with the good side down. Make certain that the workpiece is properly supported to avoid binding the blade. A *portable circular saw* is faster than a sabre saw; cut the plywood with its good side down. Be sure to use a straightedge to guide the saw for accurate cutting. On a *radial saw,* cut with the good side down; on a *table saw,* cut with the good side up. Make it your practice to fill all voids in plywood edges with a plastic wood filler (Fig. 5-1A). Sand the filled area smooth (Fig. 5-1B) and then remove all dust, dirt, and large particles before applying the adhesive.

A   B

Fig. 5-1: (A) All imperfections in the core and on the surface should be filled with wood filler. (B) After filling, sand smooth.

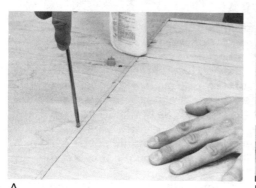

A   B

Fig. 5-2: (A) To join core pieces, use edge-rabbets. (B) Wide cleats will strengthen the joint.

Plywood that is to be covered with plastic laminate *must be absolutely clean.* Contact cement will not bond properly to dirt or oil stains, so make a careful inspection for dirt spots before starting the laminating operation. If necessary, clean up the plywood by sanding it or by using the appropriate solvent. Let the solvent dry thoroughly. When the surface is absolutely dry, you can proceed with contact cement. If a piece of plywood appears to be hopelessly soiled, do not risk delamination—set it aside for use on another non-laminated project in the future.

In most laminating projects, joinery is accomplished as described in Chapters 2 and 3. When making a long countertop, use edge-rabbets (Fig. 5-2A) to join the core pieces. Fasten with a good quality wood glue and 5/8" flathead screws. Use a wide cleat to strengthen the joint (Fig. 5-2B). Fasten it with edge strip, using wood glue and 1-1/4" ringed nails (the type used for plasterboard walls).

Cutting Laminates. Before using any contact cement, cut and match all pieces of laminate. High-pressure plastic laminate can be cut with a number of different tools. Because of the toughness of the product, carbide cutting edges are preferred—ordinary cutting edges dull too quickly. It is important that you cut from the correct side when cutting plastic laminates or the decorative surface will be chipped. Keep in mind that the cutting tool must always enter—never exit—on the decorative side. Here are the most common methods of cutting plastic laminate:

1. **Hacksaw or compass saw** (Fig. 5-3A). Fine-tooth metal-cutting blades make a neat cut; work with the decorative side up. The going will be slow, so recognize that this tool is really suitable only for short-length cuts.

2. **Power jig or sabre saw** (Fig. 5-3B). Cut with the decorative side down using a hacksaw blade (in the sabre saw). The workpiece must be adequately supported. If the work is permitted to hang in the air while you are cutting, you stand a good chance of damaging the laminate by splitting.

3. **Table saw** (Fig. 5-3C). Use a carbide-tipped blade; cut with the decorative side up. This is the best way to cut smaller, easy-to-handle pieces—larger ones can be somewhat difficult. When using a *radial saw* to cut laminates, place the decorative side down.

4. **Shears** (Fig. 5-3D). There are a number of types of shears intended especially for use on plastic laminates; with one of these you can quickly and neatly cut irregular shapes, notches, and the like. Generally, these are used by those professionals or advanced amateurs who do a great deal of laminating. How each is used depends upon its particular design. Most are equipped with carbide cutters to assure neat cuts.

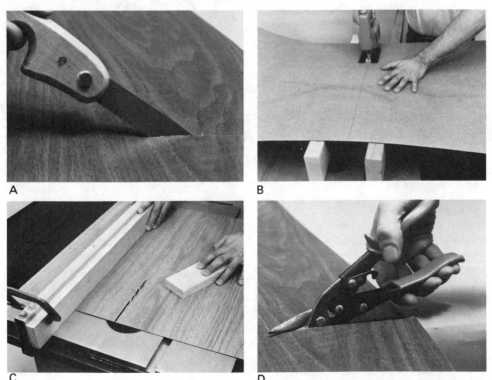

A B C D

Fig. 5-3: Cutting plastic laminates with a compass saw, sabre saw, table saw, and snips.

Fig. 5-4: How to make a perfect butt joint.

To make a perfect butt joint in laminates, overlap the two pieces and clamp a pair of wood strips to serve as guides for a router. Using a straight carbide cutter, make the cut as shown in Fig. 5-4.

Some countertops and projects require a rounded corner. To turn a radius, use a flameless heater and a jig cut to suit the curve on the core stock. The laminate is bent as the heat is applied, then allowed to cool while "clamped" in the jig.

About Contact Adhesives. It is very important that you read and follow the instructions on the labels of all materials before using them, particularly adhesives and solvents. As previously stated, both chlorinated and water-based contact cements can be brushed, rolled, or sprayed (Fig. 5-5). As mentioned earlier, solvent types are available in both brush grade (this can be troweled or rolled) and spray grade. But, regardless of type, it is wise to thoroughly stir the container contents before applying the adhesive.

Fig. 5-5: Chlorinated and water-based contact cements can be applied with a brush, roller, or spray gun.

For most brushing operations, use an animal-hair paintbrush from 2" to 4" in width, depending on the surface to be coated. Apply a *generous* coating of adhesive to both surfaces to be bonded. Do not brush too thinly since too little adhesive will result in a weak bond. Porous materials may soak up the adhesive, leaving insufficient contact cement film at the surface. If both surfaces are not completely covered, no bond will occur at the vacant area. Remember that a *uniform* wet film is desirable. Unevenness in a coating with a nonuniform layer of adhesive may result in:

1. Possible telegraphing of lumps to the finished surface.
2. Lack of proper contact between the cement surfaces, resulting in a weaker bond.
3. Thicker film spots will take longer to dry.

Recoat if dull spots appear in the dried film and the film appears nonuniform.

Fig. 5-6: Typical roller application of contact adhesive.

A carpet stipple or short-nap paint roller is good for large areas since it distributes the adhesive evenly. When using a paint roller, you can work from a tray. The foil lining shown in Fig. 5-6 speeds cleanup later. Follow the same precaution as to the thickness and uniformity of the film with a roller coating as for those applied with a brush.

The spraying method will depend upon the equipment available and the spray pattern desired. The following suggestions are for guidance, but specific details will have to be determined for each operation. It is also suggested that the equipment manufacturer be consulted. Line pressures of 60 pounds per square inch (psi) and cup pressure of 20 to 30 psi are suitable conditions for initial tests. To insure adequate coverage in critical areas, it is wise to double spray near the edges of the sheet. A medium, even pattern, using slow smooth strokes, gives best results. The air to fluid ratio should be such that good atomization occurs. Poor atomization gives slow drying, poor green strength, and less contact between the dried film. A poor spray pattern may be caused by the following:

1. A dirty air cap will often give an uneven pattern or poor atomization.

2. Too low air cap pressure will not provide adequate air flow to properly break up the stream of cement into fine globules.

3. Too high fluid or pot pressure will provide more fluid than the available air can properly break up into drops.

It must be pointed out that utmost cleanliness of equipment and proper adjustment of equipment is absolutely necessary to achieve proper spray patterns. Soak the air cap in solvent when it is not being used. Incidentally, airless spray equipment may also be used to apply contact adhesives.

Method of Applying Laminate

When applying a laminate to a core stock, the temperature should be no lower than 65° F. Similarly, relative humidity should be no less than 35% and no more than 80%. The six easy steps for applying laminate are as follows:

1. Make certain the core stock (surface to which laminate is to be bonded) is absolutely clean and smooth. Clean with solvent if necessary; sand using a belt or pad sander. At this stage, study the piece to determine the order in which you will apply laminate to the surfaces. Keep in mind that the order in which to apply various panels is determined by the edge most visible to the eye, and therefore, subject to the most abuse. An edge strip always goes on before the top piece which then neatly covers the joint (Fig. 5-7). A drawer front, on the other hand, would violate this rule and go on after the strips—including the top one—because the appearance will be better.

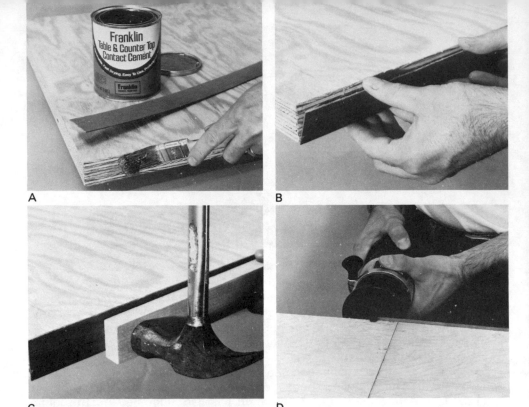

A

B

C

D

Fig. 5-7: Steps in laminating the edge: (A) Apply contact adhesive to the laminate and plywood edge. (B) Bond the strip of laminate to the edge using your fingertips to keep the surfaces apart as you go. (C) Apply the pressure immediately after the bonding, using a hammer and a clean block of hardwood. (D) Trim the excess plastic with a straight carbide cutter in a router, or use a block plane and file.

2. Once you have decided upon sequence, start by applying the contact cement to the core stock surface and its mating piece of laminate. Check the instructions on the label for drying under the right climatic conditions. To check the state of dryness, touch the surface in several places with clean kraft paper; when the adhesive no longer adheres to the paper (Fig. 5-8), the surfaces are ready for bonding.

3. Keep the mating pieces apart until they are ready for contact. Fingers alone can guide small pieces, but when bonding larger pieces, a "third-hand" helper to prevent accidental contact will be needed. There are several ways to keep the pieces apart until you are ready to bond them. One is to use a large sheet of clean kraft paper as a slip sheet; another is to use clean 3/4" diameter dowels or square sticks (Fig. 5-9) to keep the laminate and core stock separated. Align the laminate over paper or sticks; check all four sides to make sure the core stock will be covered; then, starting at one end, remove the paper or sticks and bond the laminate to the core stock.

Fig. 5-8: Surfaces are ready for bonding when the adhesive does not adhere to a piece of clean kraft paper.

Fig. 5-9: Align the laminate over the surface. As the 3/4" sticks (or dowels) are pulled out, it is bonded.

4. Most contact cements need only momentary pressure after bonding. But do not mistake momentary for light. Immediately upon bonding the laminate to the core stock, apply as much pressure as you can over the entire surface. A "J-Roller" is an inexpensive tool and is preferred to assure adequate pressure (Figs. 5-10A and B). However, a rolling pin may be used (Fig. 5-10C). You can also slide a block of clean wood about the surface and give it hefty blows with your hammer (Fig. 5-10D). Do not worry about giving the laminate too much pressure; you cannot. Just make certain you are careful when applying pressure near overhanging edges.

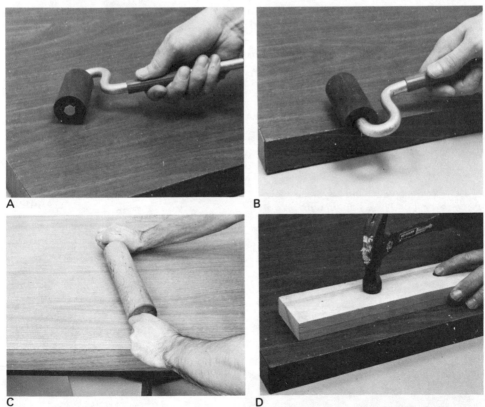

A   B

C   D

Fig. 5-10: (A) The J-roller applies great pressure. (B) Give extra attention to joints at the edges; glue lines must be absolutely tight. (C) Lacking a J-roller, a rolling pin can be used. (D) A hammer and a block of hardwood can also be used to apply pressure.

75

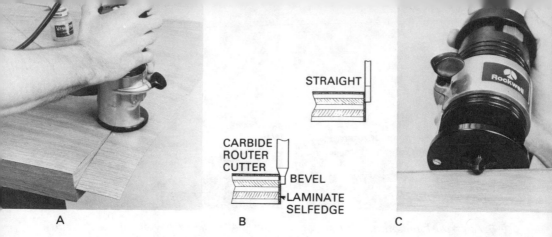

STRAIGHT

CARBIDE
ROUTER
CUTTER

BEVEL

LAMINATE
SELFEDGE

A
B
C

Fig. 5-11: Trimming a countertop: (A) Trim overhang with router with straight cutter. Protect the edge with petroleum jelly. (B) Trimming is completed when you dress the edges using a carbide bevel cutter in the router. (C) The laminate trimmer uses one cutter; it is adjustable for straight and bevel cutting.

5. The final step is trimming (Fig. 5-11). The easiest method, of course, is with a router and carbide cutter. Lacking this power tool, you will have to trim the overhang using a block plane and smooth file. Either way, the overhang is first trimmed flush (i.e., 90°) and then beveled slightly—to about 20° to 22°. Finish the beveling work with a smooth file. If a bubble should occur on the surface, it can be removed as illustrated in Fig. 5-12.

6. Cleanup. To remove contact cement remnants from the surface, use the contact cement solvent recommended on the label instructions, and use scraps of plastic laminate—*never use metal*—to scrape off the heavy globs (Fig. 5-13). Be sure to follow safety precautions for the solvent's use and provide adequate ventilation.

IRON SET FOR SILK

SINGLE LAYER OF
NEWSPAPER

CORRECTING
A BUBBLE

Fig. 5-12: To correct a bubble, place newspaper over it and press down with an iron until the heat penetrates the area. If the newspaper scorches, lower the heat setting. Roll with a J-roller or rolling pin.

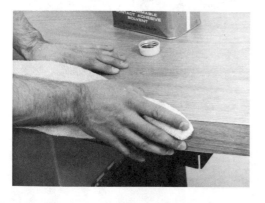

Fig. 5-13: Remove excess adhesive with a laminate scrap or a rag and contact cement solvent. Use the latter sparingly to avoid delamination at the edges.

Countertop With a Backsplash and/or Sink. Figure 5-14 illustrates a cross-sectional view of a typical countertop and backsplash setup. Actually, the backsplash is only a single or double thickness of core stock, 2" to 4" high, and the same length as the top. It is made in the same manner as the countertop. The end return strip goes on first and the backsplash horizontal strip is applied last. Use the straight cutter for all edges that will butt up against the wall or another piece of laminate, and bevel the other. After trimming the edges of the backsplash, clamp it to the countertop and fasten it with long wood screws driven up through the underside of the countertop and into the backsplash. The metal cove molding (Fig. 5-14) is optional. But if the molding is not used, run a bead of quality tub and tile caulk along the top surface before installing the backsplash.

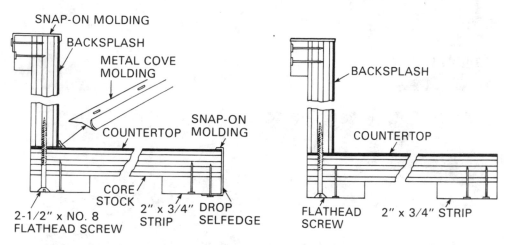

Fig. 5-14: Cross-sectional views of two styles of countertops with backsplashes.

Many backsplash surfaces are not as smooth as the plywood or particleboard countertop. For example, gypsum wallboard might have a tape joint or some spackled nail areas that, although barely visible, prohibit a smooth, even application of contact cements. In such cases a trowelable construction adhesive or a rubber based tileboard adhesive is often used to install the plastic laminate to the surface of the backsplash. Do all of your measuring, cutting, and piecing first. Trowel the adhesive onto the back of the plastic laminate, and push it into position on the backsplash.

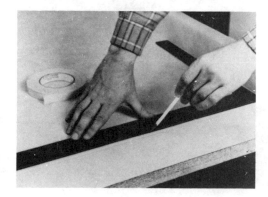

Fig. 5-15: The sink rim cutout can also be layed out by following the dimensions given by the manufacturer.

If you are going to mount a sink in the countertop, the cutout for it should be made at this point, before attaching the top to the cabinet. Locate the position of the sink and use the rim as a template to insure accurate measurement of the

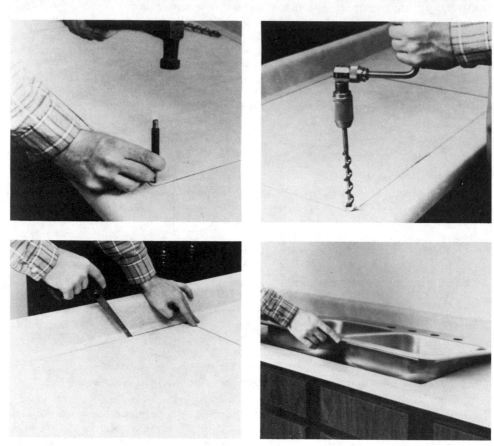

Fig. 5-16: Steps of a typical sink installation.

Fig. 5-17: Modern cocktail table (Courtesy of the Formica Corporation).

sink cutout (Fig. 5-15). The cutout should be 1/16" oversize to permit freedom for the "leg" of the sink rim. Your sink rim kit will come complete with fastening clips and corner brackets to support the sink until the clips are installed (Fig. 5-16). Nail or screw these brackets in place, caulk the inside of the rim with an acrylic or high quality tub and tile caulk, fit the rim to the sink, and set the assembly in place. Finally, install the clips around the edge of the sink according to the rim manufacturer's instructions.

LAMINATED PLASTIC PROJECTS

By following the directions and techniques outlined in this chapter, you can use plastic laminates and contact cement for many practical and imaginative applications. For example, the modern cocktail table shown in Fig. 5-17 involves the basic laminating methods covered in detail on previous pages of this chapter. The core stock material requires no wood finishing, since it will be entirely covered with decorative plastic laminate—in this case, butcherblock maple.

The box is easily assembled using butt joints at the corners. As illustrated in Fig. 5-18, the decorative laminate is veneered afterward, around the four sides in rotation. The top surface is completed before laminating the sides. By varying the dimensions, the simple cube construction can be used to make other useful furniture pieces (Fig. 5-19).

MATERIALS

Quantity	Description
2	3/4" x 16-1/2" x 22-1/2" Plywood (top and bottom)
2	3/4" x 12" x 17-1/4" Plywood (ends)
2	3/4" x 12" x 23-1/4" Plywood (sides)
4	3/4" x 3/4" x 8-3/4" Pine lumber (corner cleats)
1	17" x 23" Plastic laminate (top)
2	13" x 24" Plastic laminate (sides)
2	13" x 18" Plastic laminate (ends)
4	Platetype casters, 2" balls
-	Plastic wood putty
-	Fine sandpaper
-	1-1/4" Ringed nails
-	White glue
-	Contact adhesive
-	Adhesive solvent

Note: All necessary laminate pieces can be cut from one 36" by 72" laminate sheet. Take the time to lay out the table part sizes before cutting to assure proper yield.

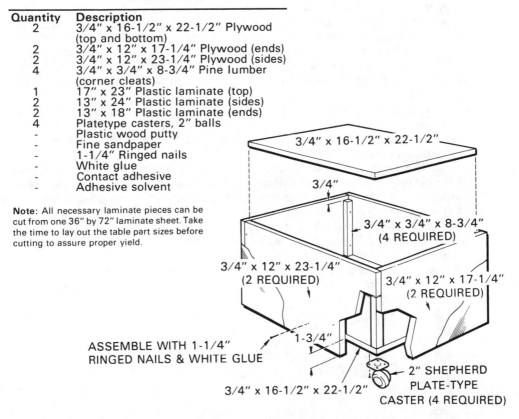

3/4" x 16-1/2" x 22-1/2"

3/4"

3/4" x 3/4" x 8-3/4"
(4 REQUIRED)

3/4" x 12" x 23-1/4"
(2 REQUIRED)

3/4" x 12" x 17-1/4"
(2 REQUIRED)

1-3/4"

ASSEMBLE WITH 1-1/4"
RINGED NAILS & WHITE GLUE

3/4" x 16-1/2" x 22-1/2"

2" SHEPHERD
PLATE-TYPE
CASTER (4 REQUIRED)

Fig. 5-18: Material list and plans for the modern cocktail table.

Fig. 5-19: The furniture cube with slight variations can also be used for other tables (Courtesy of the Formica Corporation).

There are many other projects you can do with laminated plastic and contact adhesive. For example,
- Resurface an existing flush door in color or wood grain.
- Home bars: Choose a smooth finish for easy wipe up.
- Give kitchen cabinet doors a new look. The old finish must be entirely removed from the doors before you apply the laminate; the door-reveal (around the laminate) should be the same throughout the kitchen. Since the laminate is only 1/16" thick, the molding selected for use can simply lap the joint; use miter cuts at the corners for a professional-looking job. Keep the contact cement off the exposed surfaces.
- Covering existing pantry shelves with laminate will provide an attractive, easy to clean surface.
- Built-in furniture and fixtures: The plastic surface will eliminate the periodic repainting chore.
- Resurface a tired-looking countertop. Test for looseness and thoroughly sand the old laminate before applying the new material. If the old counter material is loose or badly worn, remove it from the counter using a putty knife or scraper to get rid of adhesive and fragments of material. Then, scrape or sand the surface to leave it clean and dry. Any core material that has rotted or become delaminated should be replaced with new plywood or particleboard.
- Desktops: for a childproof surface.
- Apply the laminate to a bookshelf. Consider covering the bookcase in one color or pattern and the shelves in a second (complementary) color or pattern. Let your imagination run wild.

Installing Walls and Ceilings with an 6 Adhesive

One of the most popular and, to some degree of the word, the newest application of adhesives is in the construction field. True, adhesives were used for over 2000 years to install ceramic tile, and they have been used for years to build trailers and mobile homes, however, it has only been in recent years that they have been widely accepted for use in the construction of floors, walls, and ceilings.

In these days of high building costs, the construction industry has found that adhesives not only save time and materials, but also do the job better. Fortunately, most of these advantages of using adhesives in residential and commercial construction can also be realized by the do-it-yourselfer.

TYPES OF CONSTRUCTION ADHESIVES

There are many adhesives available to both the construction trade and the consumer (Fig. 6-1). In this chapter we will take a look at those adhesives that are primarily used to install materials on walls and ceilings, while in Chapter 7 we will cover those employed to put down various flooring products.

Construction Adhesives. General construction adhesives are one of the most versatile types of adhesives on the market today—they handle practically every job in construction, remodeling, or repair-maintenance. For instance, suggested home uses for a good construction adhesive would include the installation and repair of Acoustical tile, Bathroom fixtures (flush), brick and brick veneer, bulletin and chalk boards, Carpeting (indoor/outdoor and sponge back), cement or concrete, cement asbestos board, ceramic tile, cork, cove base, Door gaskets, door thresholds, Earthenware pottery, electric outlet boxes, Fabric, felt and fiberglass pads, fiberboard, fiberglass, flagstone, flakeboard, furring strips, Glass, gypsum wallboard, Hardboard, honeycombs (metal, paper, or plastic), Insulation, Joints, jewelry, Kick plates, Laminates (high pressure), Marble, metals, molding (metal and wood), Nameplates and signs, Ornaments, Parquet wood flooring, particleboard, plastics (certain types), plywood, pottery, Quarry tile, Railings, rattan, remodeling, Sheathing, simulated brick and stone, slate and slate blackboards, stair treads, subflooring, Tileboard, tub enclosures, Underlayment, Valances, vases, veneers, Weatherstripping, wood, wood paneling, Xmas decorating, Yawl and yacht repair, Zebrawood...just a few of many A to Z applications of construction adhesives. Always remember to read the label carefully, so that you are certain you have the correct product to suit your needs.

General construction adhesives dry exceptionally fast (10 to 30 minutes working time) at normal temperatures (above 60°F). In cold temperatures, drying will be somewhat retarded. Do not apply at temperatures below 10°F. Many formulations can be used for both interior and exterior applications. The adhesive can be gun extruded (using a caulking gun) or applied with a spatula or putty knife; some are designed for trowel applications. Their high degree of initial tack allows light pieces to be bonded with only momentary pressure. Exposure of the

Fig. 6-1: Some of the more popular "construction" types of adhesives available to both the construction industry and the do-it-yourselfer.

adhesive to air, before assembly, will increase this initial grab. The allowable open time is regulated by the temperature, humidity, and air circulation and the basic characteristics of the formulation. Most construction adhesives are thixotropic formulations that help to fill gaps and bridge minor framing irregularities. Many of the modern construction adhesives are nonflammable and nontoxic, while others contain volatile hexane, acetone, and toluene. When applying the latter, do not use near fire or flame (spark), and use only with adequate ventilation. Since some construction adhesives attack polystyrene foams, always check the manufacturer's recommendations on the container before using it on this material.

Accurate data is always difficult to prepare with regard to coverage since there are so many noncontrollable conditions to consider on cartridge applications. The speed with which the extrusion is applied, where the nozzle tip is cut, the efficiency of the mechanic or user, etc. all have a decided effect on coverage. With spatula or trowel applications from metal containers, the thickness applied, the angle of the trowel, and the depth and spacing of the notches all have an influence on coverage. Keeping in mind that a cartridge or any type of container has only so many cubic inches of area within its walls, most construction adhesives will yield similar coverages from product to product as long as the above conditions are similar and controlled. However, with reasonable care, you should expect to realize a minimum of approximately 40 to 60 square feet per gallon on trowel applications and the following scale of coverages with extrusions from cartridges:

Volume Extruded	Length of Bead Extruded with Various Bead Diameters		
	1/8 in.	1/4 in.	3/8 in.
1 U.S. Gallon (128 fluid ounces)	1,569 ft.	392 ft.	174 ft.
Large size cartridge (29 fluid ounces)	355 ft.	89 ft.	39 ft.
Tenth size cartridge (11 fluid ounces)	135 ft.	34 ft.	15 ft.

Since the above data is calculated on the basis of cubic measurements as related to fluid measurements, these figures will provide a reasonably accurate measure of volume required for most any mastic adhesive or sealant.

Most construction adhesives are available in cartridges and pint, quart, and gallon cans.

The following are general instructions for the application of a construction adhesive; specific instructions for its use are given in this chapter and Chapter 7.

1. Remove dust and other foreign matter from surfaces to be bonded.

2. There are two types of installations depending upon the type of wall surface you are working with.

　a. Studs, joists, or furring strips—Apply the proper bead (1/8" to 3/8") as determined by surfaces to be bonded (1/8" where there is contact, 1/4" for most applications, and 3/8" for bridging gaps and where the surface is irregular). A spaghetti or serpentine pattern is recommended for large contact areas, generally on 12' to 16' centers.

　b. Existing walls of plywood, plaster, plasterboard, etc.—Solid backing walls of this type generally present the best surfaces for wall covering materials. In this case, trowelable grade construction or tileboard adhesives should be used and the wall covering should be troweled over the entire surface (wall covering surface) with the recommended notched spreader.

3. Apply to one surface only.

4. Press substrates firmly in order to wet the total area to be bonded. Adhesive should spread to the edges of the desired contact area.

5. Use supplemental fasteners or weight to hold substrates in position until adhesive sets.

For cleanup, commercial products available that would be recommended are mineral spirits or lighter fluid. Avoid oily type solvents or any cleaning material that will leave an oily residue on the surface. Where the adhesive has dried, it is best to scrape all excess off with a putty knife. In some cases, soaking may be necessary for several hours to completely loosen the dried adhesive. **Note:** Always make certain that the solvent used will not mar or attack the surfaces to which it is being applied by testing an out-of-the-way area. Also, use extreme caution, as many solvents are highly flammable or combustible.

Metal Framing and Structural Adhesives. These products frequently use a neoprene base. They are designed to meet the growing demands of galvanized steel and aluminum framing members both in floor and wall systems. Many develop exceptional strength when bonding plywood, gypsum wallboard, hardboard, and paneling to these metal surfaces. Generally speaking, neoprene based adhesives have higher temperature resistance and extremely long range aging characteristics. They remain permanently resilient through wide temperature variations from -40°F to 200°F. Initial tack is quite high. The creep resistance is also unusually high on some formulations. But, remember that since a flammable solvent system is employed with these adhesives, they are flammable and their vapors can be harmful.

Panel Adhesives. These adhesives, frequently called panel and plywood adhesives or mastics, can be used to bond panels, drywall, hardboard, corkboard, bulletin boards, or chalkboards to masonry, studs, drywall, or concrete. That is, paneling of all types can be bonded tightly to studs, drywall, masonry, or furring strips with these adhesives and requires few or no nails for a smooth, one-step installation. This eliminates nail pops and patching, plus removes the danger of marring prefinished panels with hammer marks. They also overcome structural deficiencies, fill gaps, and bridge minor framing defects. Usually dispensed from cartridges, most panel adhesives will not drip from beams or sag from vertical surfaces.

While step-by-step instructions for the installation of various types of paneling are given later in this chapter, the following are the basic techniques for the use of panel adhesives:

1. Remove dust and other foreign matter from surfaces to be bonded.

2. Apply the panel adhesive to one surface only. Use a 1/8" to 1/4" bead where surface will conform closely. Use a 3/8" bead or larger on uneven surfaces where bridging is required.

3. After adhesive has been applied, position components within 20 to 30 minutes.

4. Press components firmly together to insure a good bond. Press only once. Adhesive will bridge gaps automatically. Use nails or other fasteners on areas where surfaces tend to separate. Remove fasteners after 24 hours.

5. Nail paneling at top and bottom with finishing nails.

Many panel adhesives are extremely difficult to remove after they have dried. In the wet state, a suitable solvent such as mineral spirits or lighter fluid will satisfactorily remove smears or clean up tools, etc.

Foamboard Adhesives. Plastic foams, both polystyrene and polyurethane, are being used more frequently today in all types of construction; however, they can present some problems if the incorrect adhesive is applied. This is particularly true of the common insulation varieties of polystyrene foams, which are quite susceptible to attack from certain solvents found in many panel and construction types of adhesives. Therefore, be absolutely certain that the adhesive you are using is designated by the manufacturer as being satisfactory for use on polystyrene foams. If not, it can ruin your entire installation, resulting in complete failure of your wall or ceiling system. Certain solvents, however, can be used and there are some solvent based adhesive formulations on the market that are perfectly satisfactory. Those with an emulsion (water) base are generally the best choice, since water will not bother the foam.

Polyurethane foams, on the other hand, are not too widely used for insulation in the consumer markets, particularly on do-it-yourself types of projects. They are considerably more expensive than the polystyrenes although a lesser thickness will generally provide as much insulation value as a greater thickness of polystyrene would. Currently, the largest use for polyurethanes in the home construction area is in simulated wood beams for ceilings. Polyurethane foams have a great deal of resistance to most of the solvents used in adhesives. Therefore, a good construction or panel adhesive will work nicely with the polyurethanes.

To distinguish between polystyrene and polyurethane, apply a few good "rules of thumb." Urethanes are generally of a yellowish color and have a smaller or finer pore structure. Polystyrenes, on the other hand, are generally white or blue. They may or may not have a small or fine pore structure, as the "bead" boards are merely polystyrene beads fused together.

Most adhesive manufacturers indicate on the label whether their particular formulation is satisfactory for plastic foams. Some even produce combination panel and foam adhesives that can be used on both applications. Almost all of these are very easy to use if they are designed for foam applications and will produce nearly perfect results every time. Methods of using foam adhesives are discussed later in this chapter.

Gypsum Drywall Adhesives. These adhesives are specially formulated for bonding gypsum drywall to wood or metal studs and for laminating drywall panels. They greatly increase wall strength, reduce sound transmission, and improve wall appearance overall. Those made with an elastomeric base contain no asphalt or tar and can be painted over without fear of stain-through. A high solids content makes them an excellent gap filler that bridges irregularities. Properly applied, they stay in place without sagging and allow ample time for alignment of each drywall panel. Only minimum supplemental nailing is required on walls or ceiling, thereby reducing nail pops. As a matter of fact, many so-called "nail pops," especially on ceilings, are not really pops; the nails never move. Instead, due to the cold conditions in the attic or on outside walls, condensation forms around the nail and this condensate picks up dust and dirt. Thus, what appears as a popped nail is actually a shadow of the nail head created by condensation. The use of gypsum drywall adhesives eliminates most nails and prevents nail shadows. It also does away with the necessity of the time consuming job of nail patching or spackling.

Some of the adhesives used for gypsum drywall installations are asphalt based products. There is some question as to whether these products are suitable for gypsum drywall installations. If you have any question you should find out the asphalt content from your dealer and then try to determine if the product has been given an ASTM Racking Test, Wallboard Test, or any other test that would assist you in determining quality. Should you have to use one of these products, however, due to availability or lack thereof of rubber based systems, be absolutely sure that you do not get any smears on the face surface of the gypsum board. Asphalts have a tendency to bleed through painted surfaces and are difficult to cover with ordinary paints.

Tileboard and Wallboard Adhesives. These adhesives are specifically designed for use on large sheets of prefinished hardboard, plywood, and wallboard. Smooth spreading and easy to use, tileboard adhesives form a high-strength bond to wood, masonry, poured concrete, plaster, and gypsum wallboard, and they are highly water-resistant. (Some are even waterproof.) They usually have a working or open time of approximately 30 minutes. The open time, however, may also depend upon other variables. High ambient temperatures, low humidity conditions or dry air, and high wind conditions will all shorten the working time. On the other hand, lower temperatures and high humidity conditions or moist air will tend to lengthen the working time.

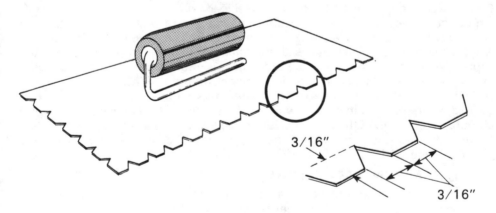

3/16"

3/16"

Fig. 6-2: Typical V-notch spreader used with tileboard adhesives.

Rubber based or elastomeric tileboard and wallboard adhesives are basically the same kind as the panel type already covered, except that they are formulated so that they can be troweled rather than extruded. On trowel applications, the adhesive is normally troweled to the back of the panel over the entire surface with a V-notched spreader having notches cut a minimum of 3/16" deep, 3/16" wide, and on 5/16" centers (Fig. 6-2). On prefinished hardboard, applications that are being installed over existing walls or gypsum wallboard, plaster, or plywood, it is recommended that the adhesive be applied over one 4' by 8' sheet of hardboard at a time. All cutting, piecing, and trimming should be done prior to applying the mastic to the back of the panel. With warped panels or out-of-plumb wall surfaces, it may be necessary to shore or brace the panels temporarily until the adhesive has taken its initial set (usually within 6 to 8 hours). On certain applications, it might be desirable to use a spot application rather than a trowel. Again, the adhesive is usually applied to the panel rather than to the wall surface. Apply the adhesive in spots or globules approximately the size of a golf ball on 1' centers.

It is not uncommon to find resin based adhesives used on tileboard installations. These mastics are characterized by their general brown to dark brown colors and extremely high viscosities. They are also generally quite "heavy." These resin systems have a great deal of cohesive strength when wet and generally do not require any bracing of the panels while the adhesive dries. However, they do not dry as rapidly as the rubber based systems and become rather hard and possibly brittle upon aging. This is particularly true when they are used in hot or warm areas such as close to a radiator or other heat source.

For cleanup operations, commercial products available that would be suitable are mineral spirits or lighter fluid. Avoid oily type solvents or any cleaning material that will leave an oily residue on the surface. Where the adhesive has dried, it is best to scrape off all excess with a putty knife. In some cases, soaking may be necessary for several hours to completely loosen the dried adhesive. **Note:** Always make certain that the cleaning solvent used will not mar or attack the surfaces to which it is being applied by testing an out-of-the-way area. Exercise extreme caution when using flammable or toxic clean-up solvents.

Ceiling Tile Adhesives. These adhesives are specially formulated for the installation of acoustical ceiling tile. They have ultra-high cohesive strength. Cohesive strength is the ability of an adhesive to adhere within itself. With ceiling tile adhesives, it is very important to have an ultra-high cohesive strength because when you install a tile on the ceiling, there is a deadweight load downward. Thus, the adhesive must be able to stick within itself to hold the ceiling tile in place without any sagging. Quick green grab is another desirable trait of a ceiling tile adhesive; it must keep the tile in place without a lot of unnecessary mechanical fasteners. Sometimes the characteristics of high cohesive strength and quick grab are desired in certain wallboard installations. Because of this, some manufacturers market these adhesives as "ceiling tile and wallboard" adhesives.

Complete step-by-step instructions for the use of ceiling tile adhesives are given on page 114.

WALL PANELING MATERIALS

There are many beautiful wall paneling materials on the market today including those made of plywood, hardboard, gypsum wallboard, and insulation board (fiberboard), as well as solid wood in various thicknesses and forms. Since many of them are prefinished, the use of adhesive makes the installation of paneling an easy, one step job.

Estimating the Number of Panels Needed

As already mentioned, most wall paneling material today is available in 4' by 8' sheets. To estimate the number of 4' by 8' panels of any material required, measure the perimeter of the room. This is merely the total of the widths of each wall in the room. Use the conversion table on the next page to figure the number of panels needed.

Perimeter	Number of 4' by 8' panels needed
36 feet	9
40 feet	10
44 feet	11
48 feet	12
52 feet	13
56 feet	14
60 feet	15
64 feet	16
68 feet	17
72 feet	18
92 feet	23

For example, if your room walls measured 14' + 14' + 16' + 16' (Fig. 6-3), this would equal 60' or 15 panels required. To allow for areas such as fireplaces, doors, windows, etc., use the following deductions:

Door   1/2 panel
Window   1/4 panel
Fireplace   1/2 panel

Thus, the actual number of panels for this room would be 13 pieces (15 pieces minus 2 total deductions). If the perimeter of the room falls in between the figures in the table, use the next highest number to determine the panels required. These figures are for rooms with 8' ceiling heights or less. For walls over 8' high, select a paneling which has V grooves and that will "stack," allowing panel grooves to line up perfectly from floor to ceiling.

Adhesive Coverage. To help you determine how much panel adhesive you will need for a given job, the following table gives the bead length in feet as related to the bead size in diameter:

	BEAD SIZE IN INCHES DIAMETER				
VOLUME	**1/8"**	**3/16"**	**1/4"**	**5/16"**	**3/8"**
	lineal feet of extruded adhesive (length)				
Small Cartridge (11 fluid ounces)	135'	60'	34'	21-1/2'	15'
Large Cartridge (29 fluid ounces)	355'	158'	89'	57'	39'
1 Gallon (128 fluid ounces)	1,569'	697'	392'	251'	174'

In round figures, 1000' of adhesive bead will require:

2/3 gallon   at 1/8" diameter bead
2-1/2 gallons at 1/4" diameter bead
5-3/4 gallons at 3/8" diameter bead

Some of the basic geometries of the more common bead patterns applied to 4' by 8' panels and the amount of panel adhesive needed are illustrated in Fig. 6-4.

Plywood Panels

Plywood panels come faced with many fine hardwoods and several textured softwoods (Fig. 6-5). Besides offering a variety of veneers and textures, plywood wall panels have the advantages of being nonsplitting, resistant to warping, and easy to install. Plywood also has fair insulating and sound-absorbing qualities and imparts structural rigidity when properly installed. Most plywood panels

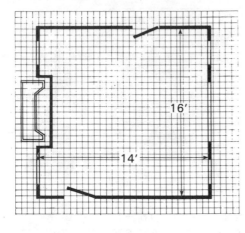

Fig. 6-3: How to figure a room for paneling.

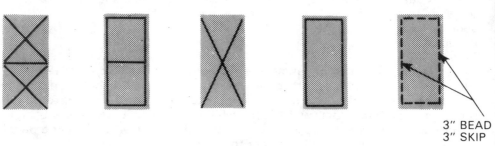

3" BEAD
3" SKIP

Fig. 6-4: Basic geometries of the more common bead patterns.

today are prefinished. Although you can save money by buying panels in their natural state, the labor of finishing is not worth the economy.

If you can, store the panels in the room for a few days before you start the job. Whether new or old, the studs should be straight, dry, plumb, and true to assure a smooth, flat wall surface. If new framing is being installed, use only No. 1 Common, thoroughly dry, straight framing lumber of uniform width and thickness. Framing should be erected on 16" centers. Where required, extra framing members should be installed to provide a nailing base for all edges of the panels. Where required, nail cats (horizontal framing members) at 4' heights for additional support of panels for every panel edge and for every 4' of panel. If you are in doubt about the dryness of the framing lumber, apply fir sticks (1/4" thick, 2-1/2" wide, and 4' long), with the grain running the short way over the face of the framing members. For studs spaced 16" on centers, use 1/4" plywood; for studs placed 24" or more on centers, apply 3/4" panels.

Plan the sequence of panels about the room so that the natural color variations form a pleasing pattern in complementary tones or in direct contrast. Hold each panel against the wall to see how it looks before you apply the adhesive.

Here are three ways to start paneling, based on individual room problems:

1. For most interiors, it is practical to start from one corner and work around the room.

2. If the wall or room has a fireplace or picture window, you should start paneling on each side (if fireplace or window goes to ceiling) or at the center (if fireplace or window does not go to ceiling), and work to the right or left around the room.

Fig. 6-5: Designed plywood arrangements (Courtesy of Weyerhaeuser Company).

3. If all the panels are the same width and window or door units are balanced across a wall area, start at the center of the wall and work both ways.

Again, do not be concerned about the natural variations in color as they will enhance the appearance of the room, as long as there is some symmetry of arrangement.

There are several tricks for laying out panels to reduce cutting as well as to achieve a pleasing pattern of joints. To avoid intricate fitting around windows and doors, start full panels on each side of the openings. On plain walls, it is best to start at the center so that fractional panels will be the same at each end. You can keep all joints vertical, the simplest arrangement, or use the tops and bottoms of windows as guidelines for horizontal joints.

Be sure that the panels are square with the adjacent wall (at corners) and ceiling before installing. If the panel is not square with the adjacent wall, scribe it to the corner. Keep the bottom of each wall panel about 1/4" above the floor to allow space for the lever used to pry the panel tightly against the ceiling. As the panels go up, keep checking them for plumbness. Keep a level handy for truing up and down (vertical position). Molding will take care of the irregular meeting with floor and ceiling.

When cutting plywood with a handsaw or on a table saw, it should be cut with the good face up. If you are using a portable electric handsaw, either circular or sabre, cut the plywood with the good face down. If you are using a radial saw, cut the plywood with the good face up for crosscuts and miters, and down for ripping. With a handsaw, or on a table saw, permit only the teeth of the blade to protrude through the work. For smooth cuts, use blades that have teeth with no set and that are hollow ground. Special small-toothed blades are available for cutting plywood.

89

Installing With Adhesive. The application of plywood paneling with panel adhesive is widely employed by do-it-yourselfers. Its use largely eliminates the need for brads or nails and the resulting concealment of their heads. If an old wall is in good condition, smooth and true, the adhesive can be applied directly to the back of the panel all around the edges in intermittent beads about 3" long and spaced about 3" apart (Fig. 6-4). Keep the adhesive at least 1/4" from the edges of the panel and be sure that it is continuous at the corners and around openings for electrical outlets and switches. Additional adhesive should be applied to the back of the panel in horizontal lines of intermittent beads spaced approximately 12" to 16" apart. Once the adhesive is applied, the panel may be pressed against the wall. It may be moved as much as is required for satisfactory adjustment. To make this easier, drive three or four small finishing nails about half their length through the panel near the top edge. The panel can then be pulled away from the wall at the bottom with the nails acting as a hinge. After any adjustment has been made, a paddle block should be used to keep the panel pressed back on the wall, and then the nails are driven home. (These will be covered by a molding.) A rubber mallet or a hammer and padded block should be used on the face of the panel to assure good adhesion between the panel and wall. Never attempt to apply adhesives on plaster walls in poor condition, with flaking paint or wallpaper that is not tightly glued. If the plaster seems hard and firm and does not crumble when you drive a nail into it, it is probably safe for adhesives.

There is an alternate method of installation on existing wall surfaces that provides you with an outstanding paneling job. Where the existing wall is plaster, gypsum board, or another smooth backing, a trowel grade construction adhesive or tileboard adhesive may be used. In this case, trowel the mastic over the back surface of the panels utilizing a trowel with notches approximately 3/16" deep, 3/16" wide, on 5/16" centers, unless otherwise specified by the manufacturer (see page 85).

Adhesives may also be used on furring strips and open studs. It is applied directly to each furring strip or stud in continuous or intermittent beads (Fig. 6-6). But, before applying the panel adhesive to open studs, it is a good idea to inspect all studs for low spots with a chalk line or straightedge. Mark those studs that have one or more low spots. (To locate studs after paneling is up, put chalk mark on floor at base of stud.) Draw a plumb line on the stud that is 4' from the corner. Where studs are straight, apply a 1/4" high bead of panel adhesive.

Fig. 6-6: Applying panel adhesive to studs.

Where studs have one or more low spots, apply a 3/8" bead. On the studs at the edges of the paneling, apply the adhesive inside the plumb line to keep the adhesive from squeezing out at the edges. On the other studs, follow the plumb line.

To install paneling on metal studs, proceed as above; however, be sure that the adhesive manufacturer recommends it for metal studs. Employ self-tapping screws rather than nails to hold panels in place at the top and bottom.

When using furring strips, be sure the surface is smooth or only slightly irregular. Knock off all high spots. With furring strip backing, the adhesive should be applied to the strips.

Regardless of backing—solid, studs, or furring strips—the first panel you apply is the most important. Always start in a corner. Usually, the first corner you see as you enter the room is the ideal starting place. When installing the first panel, you must be sure that the panel is perfectly plumb (that is, the outer edge is exactly perpendicular to the floor and ceiling). If the outer edge of the panel does not fall on a stud or furring strip, cut the panel to fit. With the first panel plumb and on a stud or furring strip, the rest of the panel edges will land on studs or furring strips. This, of course, need not be considered if paneling is being installed over a solid backboard or to an existing wall surface.

If the corner is not even, you can do the following: Tap two finishing nails into two grooves at the top edge of the panel, just protruding through the back of the panel. Place the panel in the corner at a distance that can be spanned by your scribing compass. Using a carpenter's level, plumb the panel on the outer edge. When plumb, drive the nails partially into the stud or furring strip. Then, using your scribing compass, scribe a line from top to bottom (Fig. 6-7). For a good line, use a china marking pencil. Cut the scribe line with a finishing saw. This method can also be used for corners with fireplaces, bricks, stone, etc.

Angle the first panel toward the corner plumb line, align the panel, and press the panel just long enough to spread the adhesive (Fig. 6-8). Panels covering studs with low spots will release from the low spots for a smooth wall installation. Although the panel may be moved for slight adjustments, do not press on

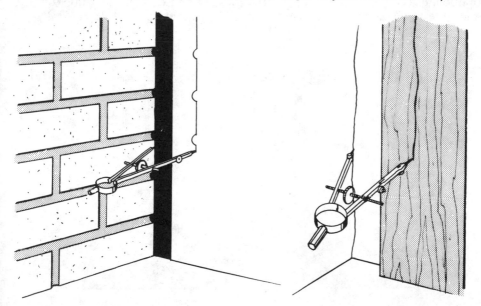

Fig. 6-7: Method of scribing irregularities on a wall.

Fig. 6-8: Installing the first panel.

the panel after the first positioning pressure. The second piece of paneling should fit snugly to the edge of the first piece. Keep all subsequent panels plumb, trimming the top or bottom of the panel, if necessary, due to unevenness in the ceiling or floor. As mentioned earlier, use three or four finishing nails at the top and bottom—under the area to be covered by molding. If any butt edges bow out, drive a finishing nail through a small block of wood and through the joint. Remove after 24 hours.

Measure from the edge of the last panel installed to the edge of the untrimmed door or window opening. Mark this on the panel to be cut. Next, measure the height of the door or window and mark this on the panel (Fig. 6-9). Cut to fit, remembering the side from which to saw the panel. If it is possible, your cut out panels should meet at the middle area above and below your window or door. The panel will be easier to work with than if the cut is out of the middle of the panel. If you must cut out of the middle of the panel, be sure your measurements are correct. Drill a 3/4" hole (from face side) at the corners of your measurements to give you a turning corner if using a keyhole or sabre saw for the cutting. In a remodeling job, you should fur out the window and door frames to equal the panel thickness, so that your window and door moldings will fit naturally.

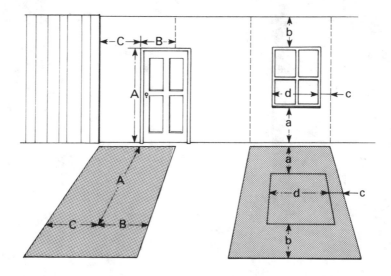

Fig. 6-9: Measuring for doors and windows.

There are several ways to mark the electrical outlets. The two most common are: (1) After fitting the panel correctly, place a soft wood block over the panel at the outlet. Tap the block sharply enough for the outlet to indent the back side of the panel; or, (2) rub chalk on the outlet, fit the panel, and tap lightly. The chalk will mark the back of the panel. Drill small holes at the corners of the outlet marks from the back of the panel. From the face of the panel, drill larger holes and saw the outlet hole from the face side of the panel with a key hole saw.

Just as a fine painting requires the right frame, prefinished paneling needs decorative moldings for a professional appearance (Fig. 6-10). Moldings frame the doors and windows, cover the seams and joints at the ceiling and floor, and finish off and protect corners. Most of the larger companies make hardwood moldings that will harmonize with their various species of plywood. Some sell moldings that are already finished to match, while others sell special stains that will enable you to blend the trim in with the finished wall later on. Some companies also make veneered metal moldings with a matching wood facing to conceal exposed plywood edges and to finish off inside or outside corners.

Wood moldings should be cut with a fine-toothed saw. To splice lengths of moldings along a wall, 45° cuts are made at the same angle on both pieces. Where moldings meet at right angles—corners, around windows and doors—use a miter box for accuracy. Trim both pieces at opposite 45° angles so that together they form a tight right angle. When installing molding, apply a 1/4" bead of panel adhesive to the backs of moldings and press it into place at the desired positions (Fig. 6-11). Masking tape and strips placed perpendicularly over the moldings will help hold them in position until the adhesive sets.

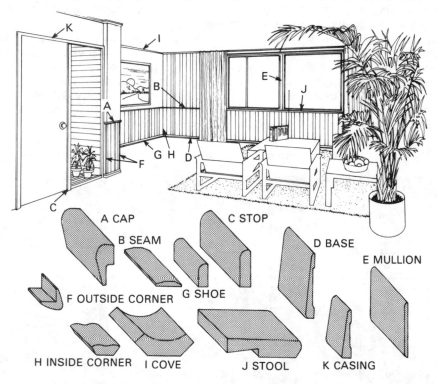

Fig. 6-10: Various wood moldings and how they can be employed.

*Fig. 6-11: Installing molding
with an adhesive.*

Hardboard Panels

Hardboard specially manufactured for use as prefinished paneling is specifically treated for resistance to stains, scrubbing, and moisture. It is also highly resistant to dents, mars, and scuffs. In most cases, the material is prefinished in wood grains, such as walnut, cherry, birch, oak, teak, and pecan, and in a variety of shades. It may be smooth-surfaced or random-grooved (Fig. 6-12). In addition, there are the decorative and work-saving plastic-surfaced hardboards which resist water, stains, and household chemicals exceptionally well. A typical surface consists of baked-on plastic. Most hardboard is sufficiently dense and moisture-resistant for use in bathrooms, kitchens, and laundry rooms. The variety of finishes and sizes is extensive. Finishes include rich-looking wood grains, exceptional marble reproductions, plain colors, speckled colors, simulated tile, lace prints, wallpaper textures, and murals. Vinyl-clad panels are also available in decorative and wood grain finishes.

Measure for the panels you will need. It always helps to measure and then remeasure the entire area to be paneled, and then plot these measurements to scale on graph paper. Make a plan view and also a layout of each wall showing fixtures, electrical outlets, and other details. Also indicate each type of molding and where it will be installed. Remember, panels come in 4' by 8' sheets and most moldings in 8' lengths. Also keep in mind that the information given on page 86 holds good for determining the number of 4' by 8' panels of any material.

Fig. 6-12: Typical hardboard installations (Courtesy of Weyerhaeuser Company).

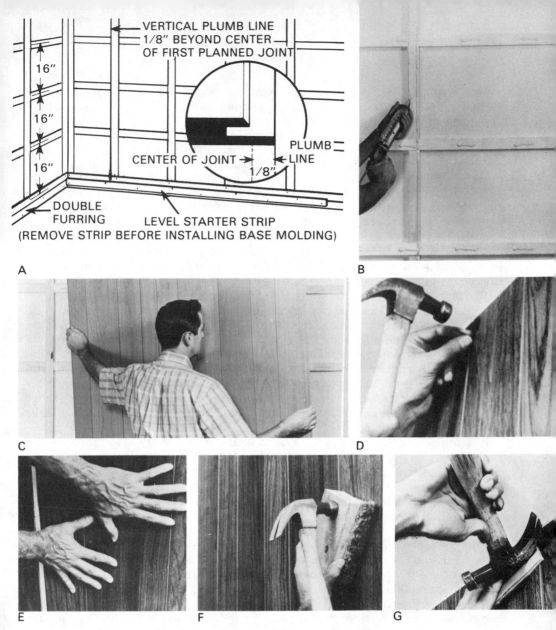

Fig. 6-13: Applying hardboard on furring in new or old construction. (A) Nail 1" by 2" furring strips horizontally into old wall at stud locations, spaced 16" apart. Apply vertical furring where panel edges are to be bound. (Furring may be attached vertically with a cartridge-type adhesive, if desired.) (B) After making sure that all surfaces to which adhesive is to be applied are clean and dry, apply 1/8" thick continuous ribbon to furring or other surfaces to which panel edges are to be bonded. Apply intermittent ribbon (3" bead—6" open space) to intermediate furring. Adhesive and room temperatures should be between 60 and 100°F during application. (C) Move panel into position over furring strips and immediately press into position. (D) Install two nails at top of the panel to maintain its position, leaving the heads exposed for subsequent easy removal. (E) With uniform hand pressure, press the panels firmly into contact with the adhesive bead. (F) After fifteen to twenty minutes, reapply pressure to all areas to be bonded, using a padded block of wood and a hammer or mallet. A final set is thus provided. (G) Carefully remove the nails, protecting the panel surface with a scrap of carpeting. When installing base, follow the procedure shown in D. If prefinished plastic moldings are not used, a wood molding can be stained or painted to harmonize or contrast with the paneling (Courtesy of Masonite Corporation).

Use 1/4" or 5/16" hardboard panels over open framing. All panel edges should be backed by a stud, furring strip, or solid wall. Studs or framing members should be spaced no more than 16" on center. Utilize 1/4" or 5/16" board thicknesses for structural wall members. Hardboards 1/8" and 1/4" thick should be applied over solid backing. Quarter-inch-thick boards may be applied directly over studding or stripping not over 16" on center. For most new construction, however, it is recommended that 1" by 3" furring strips be glued to the studs. To bring out the face of the strips to a level plane, shim the furring with ordinary shingles driven between the wall and strip. Glue the shingles in place with a construction type of adhesive. When furring over masonry, apply the strips with construction adhesive. Arrange the hardboard panels around the room in the desired sequence, standing them against the wall. Do not slide the panels over each other. As shown in Fig. 6-13, 1/4" hardboard panels are applied over furring strips in the same manner as for plywood.

To prepare an existing solid wall for the hardboard panels, remove all wallpaper, scaly paint, or dirt. Then, remove such fixtures as the lavatory, toilet tank, and wall-hung accessories. For a wainscot installation, strike a level line extending horizontally around all walls at the predetermined wainscot height. Plan for a gap of 1/4" between the paneling and the floor to be covered later by base trim. Measuring up from the floor, draw a line around the room to correspond with the top of the base trim. Allow a total of 3/16" between panels, to make provision for a 1/16" expansion space between each panel edge and the inside of the divider molding. *Never force panels into grooves of the metal moldings and always leave a little space for expansion.*

It is important that the work be done in the proper sequence. Some of the metal moldings (Fig. 6-14) must be installed before the wall panels are fitted and placed in position, but it is convenient to install others as the job progresses. Remember that the moldings overlap the panel edges, therefore do not fit molding over the last two panel edges until the panel has been positioned properly. To cut the metal molding, use a hacksaw and miter box, and file off the rough edges after cutting. The molding has wide flanges through which nails are driven; panels conceal nailheads. Where moldings meet in a corner, flanges must be cut back so they do not overlap. Some panels have special corner moldings available. With planning, you can make a decorative feature out of the molding lines.

Begin at an inside corner; nail cap molding along the line marking the top of the wainscot (use miter joints in corners). Also nail an inside corner molding in the corner, extending from the top of the base trim to the bottom of the exposed flange of the cap molding. It will be necessary to cut away a little of the wall flange where it overlaps the cap molding. Next, check the corner to see if it is straight and plumb. If not, it will be necessary to scribe the edge of the panel and

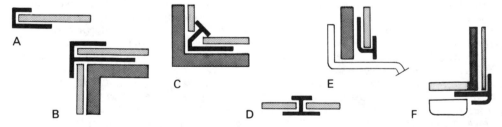

Fig. 6-14: It is essential that the proper moldings be used. These may be either aluminum or plastic of the following types: (A) cap molding; (B) outside corner; (C) inside corner; (D) divider; (E) tub molding; and (F) cap molding with slots cut.

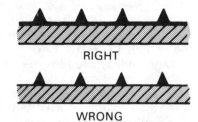

RIGHT

WRONG

Fig. 6-15: (Top) Correct amount of adhesive. (bottom) Too little adhesive.

trim it to conform to the corner. Fit, *but do not install*, the divider molding for the exposed vertical edge of the panel (remove part of wall flange where it overlaps the cap molding).

For full-height panels, start at an inside corner and nail a strip of inside corner molding extending from the top of the base trim to the ceiling. Next, install a piece of cap molding along the wall/ceiling intersection. Place a panel in position and, if necessary, scribe and trim the panel edge to fit the corner. (Make sure the edge away from the corner is plumb.) Cut and fit a divider molding, but do not install it yet.

Panels are cut face up using a sharp crosscut saw, 8 to 12 teeth per inch. If an electric portable handsaw is used, the panels should be face down when cut. Take care not to scratch the plastic surface when handling and cutting the panels.

After the panel has been fitted, lay it face down on the padding (to prevent scratching) and wipe off any dust on the backside. Using a notched trowel and an adhesive recommended by your building materials dealer, spread adhesive over the entire back of the panel (Fig. 6-15). If not otherwise specified, a trowel with 3/16" notches is recommended (Fig. 6-2). For a waterproof seal, put adhesive or caulking bead (as recommended by the adhesive manufacturer) in the molding grooves. Slip the panel into position and press it tightly against the wall. Next, apply adhesive in the groove of one of the divider moldings (previously fitted and set aside) and slip it into place along the panel edge. Check to see that the divider is plumb, then fasten that edge of the panel by nailing through the exposed flange. Return to the panel and press it firmly against the backing, working from the center toward the edges. Repeat this after 20 minutes to ensure good contact. Also, remove any excess adhesive from the finished surface as soon as possible with a soft cloth and mineral spirits or turpentine. Be sure the room is well ventilated to dissipate the solvent, and do not permit smoking or open flame in the room while solvent is being used.

Install additional panels and molding in your preplanned sequence. To put a prefitted panel into place when only one side is open, i.e., surrounded by moldings on three edges: Bend the panel into a curve until the two opposite edges can be slipped into the moldings; then release the curve and slide the panel edge into position behind the third molding. Finish the job by installing the base trim at the bottom of the wall.

Before starting to panel walls around bathtubs or showers, turn off the water supply and line the tub with protective material to avoid scratches. Then remove towel bars, soap dishes, faucets, etc.

The joints between the tub and paneling must be made permanently watertight. First, bend flexible tub molding to fit closely against the tub/wall intersection. Clean this area carefully to ensure a good bond, then caulk the back of the tub molding and nail it in place. Seal the nail heads with caulking.

It is best to start with the panel at the faucet end of the tub. Depending on whether you want a wainscot or a full wall installation, follow the general procedures given previously. Consider, however, if the best procedure will be to start by installing a molding at the inside corner and work both ways or to start at the outside corner and work progressively around the tub. Horizontal joints must not be installed in shower areas.

Using heavy paper, make a template (Fig. 6-16) showing the location of faucets, valves, shower head, etc. After double checking the template for accuracy, transfer the pattern to a panel that has already been cut to fit the wall. To make the cutouts, drill a starter hole, then use a keyhole saw to cut the opening. When the panel has been fitted properly, apply adhesive to the panel back, fill the tub molding groove and other moldings with caulking for a waterproof seal, and press the panel into position. When continuing with the installation, be sure that the panel edges, joints, and moldings are well sealed with caulking for a watertight installation.

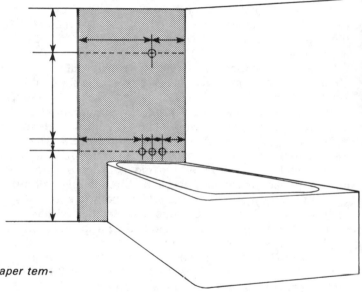

Fig. 6-16: Making a paper template for bath tub.

Fit panels loosely around windows, doors, and other obstructions and at the wall-ceiling intersection to permit normal expansion and contraction. To locate electrical outlets, chalk the edge of the box and carefully fit the panel in place over the chalked box. Strike the face of the panel several times with the heel of your hand to transfer the chalk outline to the back of the panel. Check the location by measuring from the edge of the adjoining panel and down from the ceiling or the top of the wainscot. Drill pilot holes from the back, then with the face side up, cut the molding with a keyhole saw.

A cap molding can be made adjustable by nailing through the slots in the flange instead of the holes. The slots are saw cuts made as shown in Fig. 6-14. Place the nail at the end of the slot (do not drive it too tightly) to permit the molding to slide under the molding head. Move the molding 1/4" to the side until after the panel is in position. Then, place a wood block along the side and gently tap the molding into place with the flange covering the panel edge. Short lengths of cap molding can be attached to the panel edge with adhesive or caulking before the panel is put into place on the wall. Be sure to put adhesive on the back of the molding as well as the back of the panel. If necessary, temporary bracing may be used to hold the panels into position until the adhesive has set. Usually, this is not necessary, as the moldings will prevent the panels from moving. An installation can be carried around an outside corner and finished off with a combination of moldings as shown in Fig. 6-14. Variations of this treatment can be adapted to other problem situations.

Gypsum Wallboard

Gypsum wallboard—also called plasterboard, gypsum-board, wallboard—is low cost, durable, and easy to handle when installing. This wallboard is a sheet of material composed of a gypsum filler faced with paper. Sheets are normally 4' wide and 8' in length, but can be obtained in lengths up to 16'. The edges along the length are usually tapered, although some types are tapered on all edges. This allows for a filled and taped joint. This material may also be obtained with a foil back, which serves as a vapor barrier on exterior walls. It is also available with vinyl or other prefinished or predecorated surfaces. In new construction, a 1/2" thickness is recommended for single-layer application. In laminated two-ply applications, two 3/8" thick sheets are used. The 3/8" thickness, while considered minimum for 16" stud spacing in single-layer applications, is normally specified for repair or remodeling work and most do-it-yourself projects.

When the single-layer is used, the 4' wide gypsum sheets are usually applied vertically, but can be installed horizontally (Fig. 6-17) on the walls after the ceiling has been covered. The horizontal method of application is best adapted to rooms in which full-length sheets can be used, as it minimizes the number of vertical joints. Also, plan to work from the top of the room down, i.e., the ceiling and/or top half of the walls, in that order. On walls, this approach provides a better joint, and panels already installed are less likely to be marred by other panels being handled. Where joints are necessary, they should be made at windows or doors. End joints over openings should be on studs. When ceiling heights are over 8' 3", or when this horizontal method results in more waste or joint treatment, the vertical method of application should be used.

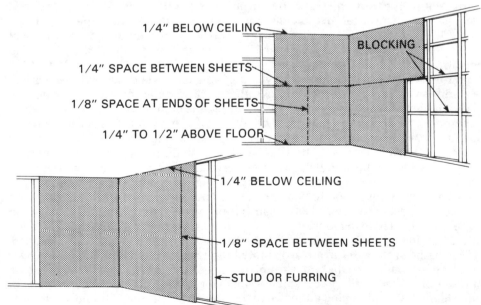

Fig. 6-17: (Top) Horizontal and (bottom) vertical installation of gypsum board.

Today, the panel adhesive method of installing gypsum paneling is rapidly replacing nails, especially by the do-it-yourselfer. And the reasons are very simple:

1. It reduces the possibility of nail popping.
2. It prevents the dirty nail marks caused by condensation.
3. It eliminates the time-consuming job of spackling or treating the nailheads. When installing predecorated or finished gypsum boards, the use of panel adhesive is almost a "must."

When using an adhesive, however, the panels should be preconditioned by a simple "bowing" procedure. This assures pressure at the center for maximum contact while the adhesive sets. To accomplish this panel bow, cut the panels (cutting details are given later in the chapter) to the proper ceiling height (1/4" to 1/2" shorter than floor-to-ceiling height, to compensate for uneven ceilings and avoid having to force panels into position). Then, stack the panels as shown in Fig. 6-18, using 4" high wood blocks to create the "bow." Two 2 by 4's running across the width of the stack are ideal. Stack with the finished face down. Pad the wood block with carpeting or similar material to prevent damage to the finished surface. Proper bowing can take from one to several days, depending on humidity conditions.

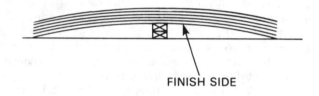

FINISH SIDE

Fig. 6-18: Method of bowing gypsum panels.

Installing Gypsum Panels. If the surface is smooth, very little preparation is necessary except for removing moldings and loose spots in wallpaper or plaster. Apply 1/4" beads of an adhesive to the back side of the gypsum board in vertical strips, 12" to 16" apart, and 3/4" to 1" in from each edge. Place the panel into position. With a board scrap, block the bottom of the panel up to 1/4" to 1/2" from the floor. Press the panel firmly to the wall using a sliding motion to spread the adhesive. Nail at the top. Remove the block and nail across the bottom.

When paneling a room with rough plaster walls or masonry walls (generally a basement), furring strips must first be attached to the wall. The vertical furring strip that will support the starting edge of the starting panel is butted snugly against the adjoining wall. The first panel is also butted against the adjoining wall. The width of this panel determines the location of the centerline of the second vertical furring strip. Each following vertical strip should be spaced so that the panel edges cover one half of each strip. The centers of each horizontal strip should be spaced 16" apart, with a strip framing each door and window. When panels are applied to masonry walls below grade, the use of a vapor barrier, such as plastic, foil, or a sealant, between the wall and the paneling is recommended.

This same system of furring strip installation is also used when paneling over very uneven walls or walls with a rough textured surface. In order to make a level surface or plumb wall, place small shims under the furring strips whenever necessary. Before applying adhesive to the furring strips, they should be cleaned free of the dust, oil, etc. Remember, for proper bonding of the adhesive to any surface, it must always be clean.

Gypsum panels may also be applied directly to the studding. Recommended stud spacing is 16" on centers. Studs where panel edges are to be bonded receive two continuous 1/4" beads of adhesive, one near each edge of the stud face. Intermediate studs receive *one* continuous 1/4" bead of adhesive approximately centered on the stud face. After pressing the panel into position, nail the panel along the top and bottom ends. These nails will later be covered by the base and cove molding. The bottom of the panel should be about 1/4" off the floor. With uniform hand pressure, press the panel into full contact with the adhesive bead. Adjacent panels should make only moderate contact and never be forced together.

Ceilings of gypsum board go up first, of course. Remember that the panels should be installed at right angles to the joists. The gypsum boards themselves can be installed in the same manner as for walls. But, holding the ceiling board in position can be difficult. If you are working alone, you can solve this problem by making a "T" brace of a two foot long 1" by 4" nailed to the end of a 2" by 4" of sufficient length to reach from the floor to the height of the ceiling. The T-brace can be used to hold the panel until the adhesive sets. Nails should also be used to support a panel while the adhesive dries. On ceiling installations, nailing should be spaced 6" to 8" around the perimeter of the gypsum wallboard. Two or three nails should be used in the center (field) of the board. These center nails can either be removed or driven through the gypsum board, and the holes can be spackled after 24 hours.

Taping Joints. With your room completely walled, you can now begin the final stages of the installation, which consist of taping and cementing joints, filling hammerhead dimples, and smoothing over any possible surface irregularities. Joint cement, "spackle," is used to apply the tape over the tapered edge joints and to smooth and level the surface. It comes in powder form and is mixed with water to a soft putty consistency so that it can be easily spread with a trowel or putty knife. It can also be obtained in premixed form. The general procedure for taping is as follows:

1. Use a wide (5") spackling knife and spread the cement in the tapered edges, starting at the top of the wall (Fig. 6-19A).

2. Press the tape into the recess with the putty knife until the joint cement is forced through the perforations (Fig. 6-19B).

3. Cover the tape with additional cement, feathering the outer edges (Fig. 6-19C).

4. Allow the cement to dry, sand the joint lightly, and then apply the second coat, feathering the edges (Fig. 6-19D). A steel trowel is sometimes used in applying the second coat. For best results, a third coat may be applied, feathering beyond the second coat.

5. After the joint cement is dry, sand smooth the area. (An electric hand pad sander works well.)

Interior corners between walls and ceilings may also be concealed with some type of molding. When moldings are used, taping this joint is not necessary. Wallboard corner beads at exterior corners will prevent damage to the gypsum board. They are fastened in place and covered with the joint cement.

Basement Installation. Thanks to modern adhesives and science, it is a simple task to insulate and finish the walls of your basement or other rooms in your house. Use plastic foam insulation (either polyurethane or polystyrene) and, although the use of gypsum board is discussed, any of the paneling material already covered can be used, if your local building or fire code permits.

Before insulating the walls, be sure the surface to which you will be bonding is

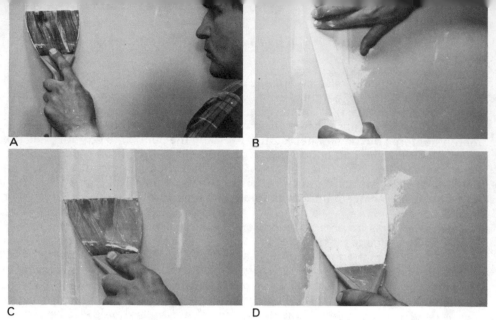

A

B

C

D

Fig. 6-19: How to tape joints.

structurally sound, clean and dry, and is free of grease and loose paint. With poured concrete foundations, it is especially important that the surface is free of any form release substances that may have been used when the foundation walls were poured. Since these materials usually contain oils, greases, and/or silicones, the masonry surface should be washed down thoroughly with tri-sodium-phosphate (TSP) or a strong detergent.

With masonry nails or construction adhesive, attach horizontal wood strips (2" wide and the same thickness as the foam insulation) continuously along the top and bottom edges of the masonry walls and around windows and doors. Cut the foam insulation to fit around any surface projections, such as windows, wood strips, electrical outlets, and conduits. This is accomplished easily with any sharp utility knife by scoring the insulation and snapping it, or just by cutting all the way through it. **Caution:** Before applying any adhesive on polystyrene type foams, make certain that the adhesive is formulated for this purpose. Polystyrene foams can be severely damaged by certain solvents contained in adhesive formulations.

Following the instructions on the cartridge, apply a continuous bead of foam adhesive around the perimeter and strips in the center of the foam (Fig. 6-20). Or, apply daubs the size of a golf ball on 12" centers around the perimeter and through the field of the foam board. Do one board of insulation at a time. When the adhesive has been applied, place the foam against the wall surface horizontally or vertically, whichever is more practical. Uniformly press the insulation board to the wall surface over every square foot to assure a positive and intimate bond. Make any necessary adjustment. Repeat this action for every board of foam insulation. Allow the adhesive to dry 24 hours before covering the foam with any other materials. However, again make sure that the adhesive you use to install the paneling is compatible with and will not damage the foam.

Do not insulate over water or drain pipes. Instead, butt the insulation up to the pipes and, if possible, wedge some foam behind the pipes. During the winter months, heat from the interior of the house may be necessary to prevent pipes from freezing and bursting. Bridging the pipes with paneling is acceptable.

When all the surfaces to be insulated have been covered with insulation, you are ready to apply some type of gypsum wall covering. This should be cut to fit the entire height of the wall, since it will fit over the foam and the wood strips. Cut out any openings for electrical outlets and switches as needed. Then, apply foam adhesive to the wall covering in the same manner as you did to the insulation. Stand the gypsum board vertically and adhere it to the foam, firmly and uniformly applying pressure over each square foot of surface. Use finishing nails at the top and bottom to fasten the panels in position until the adhesive sets. Then, finish off the wall covering with molding.

If you do not wish to insulate your basement walls, the paneling can be installed on furring strips; never apply paneling directly to masonry walls. The furring strips may fasten with a construction adhesive and/or concrete nails or nail anchors. Shim the strips where necessary. The paneling itself is bonded to the furring with panel adhesive, as previously described.

You can also build stud framing for paneling over masonry walls. Insulation between the studs is recommended for maximum comfort. Use a foil moisture barrier or glass fiber insulation with a vapor barrier sheet, applying so the vapor barrier faces you. The panel adhesive is applied to the studs.

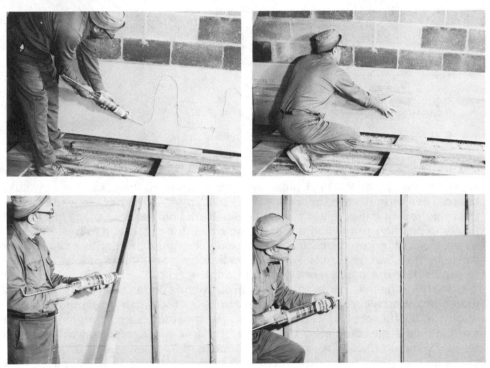

Fig. 6-20: Applying adhesive to foam.

Patching Gypsum Panels. Small dents or cracks in a gypsum board panel can be repaired with joint cement. But, patching large holes requires some type of supporting brace held in place with adhesive. As shown in Fig. 6-21A, square the hole in the panel and cut the damaged portion. Do not attempt to push or snap the broken gypsum board back into the hole; make the cutout large enough so the patch will surround sound, undamaged gypsum board material. Then, cut two pieces of 2" by 4" to a length of approximately 8" longer than the distance across the hole. Apply panel adhesive (or white glue) to one piece of the 2" by 4", then insert it through the hole and tie it to another piece of 2" by 4" which is held parallel to it (Fig. 6-21B) but in front of the wallboard.

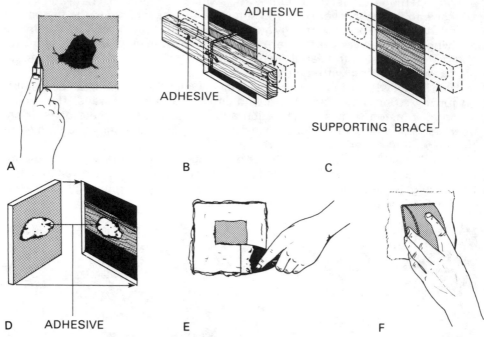

Fig. 6-21: Method of fixing a large hole in gypsum board.

Allow the two pieces of 2" by 4" to remain tied in this position until the adhesive dries and holds the back piece of 2" by 4" firmly to the back of the gypsum board. Then, remove the supporting piece of 2" by 4" in front of the wallboard by untying the string (Fig. 6-21C). The adhesive will hold the back piece of 2" by 4" firmly in position to provide a supporting brace for the wall patch. Now, cut a patch block to fit into the cutout area. The patch should be slightly smaller than the hole itself, but it should be cut to fit as tightly as possible. Apply an adhesive to the back of the patch block and fit it into the hole (Fig. 6-21D).

Once the adhesive has dried, use a firm putty knife or patching spatula to apply joint compound all around the patch board. Work the patching compound thoroughly into all cracks (Fig. 6-21E), then scrape away any surplus material and let the patched area completely dry. When the area has completely dried, use a regular sanding block (Fig. 6-21F) and a piece of fine sandpaper to sand away any high areas on the patched surface. A prime coat then can be applied to get the wall ready for painting.

Solid Wood Paneling

The use of adhesive with solid wood panels eliminates the need for puttying nails and repairing hammer marks—and this can be very time consuming when installing solid wood paneling. With solid panels, various types and patterns of woods are available for application on walls to obtain the desired decorative effects. For informal treatment, knotty pine, redwood, whitepocket Douglas fir, sound wormy chestnut, and pecky cypress, finished natural or stained and varnished, may be used to cover one or more sides of a room. In addition, there are such desirable hardwoods as red oak, pecan, elm, walnut, white oak, and cherry also available for wall paneling. As mentioned earlier, most types of paneling come in thicknesses from 3/8" to 3/4"; widths vary from 4" to 8", lengths from 3' to 10'.

To estimate quantities of board paneling, measure and multiply the areas to be paneled. Deduct for major openings only. Then add 10 to 20% for waste in fitting, lap of boards, if any, and the difference of rough width from finished width.

Solid wood is subject to shrinkage and swelling, even though kiln-dried. After delivery, therefore, stack the lumber inside the house at a temperature as close to room temperature as possible. The paneling should never be stored where it will be exposed to weather or to excessive moisture. The building or room in which the wood planking is to be installed should be completely closed in and dry before installation begins. Masonry and other work involving moisture should be completed and dried. A moisture barrier such as polyethylene plastic sheeting (4 mils thick) should be provided behind paneling where any danger of moisture penetration exists. It must, however, be placed between the original wall surface and the furring strip, as an adhesive will very rarely develop a satisfactory bond on polyethylene sheet. This is a requirement on outside walls and on all concrete or masonry walls. Also, if you intend to install paneling over a masonry wall that is often damp, it is a good idea to apply a wood preservative containing pentachlorophenol to the back of each panel. This will protect the paneling against moisture, mildew, fungus, and termites and other insects.

Before doing any installation, lay out the boards on the floor adjacent to the installation wall. Arrange the most attractive combination of widths, lengths, grain patterns, and shades of color. Then install the boards in the selected sequence. Use shorter pieces at the top and bottom of the wall area, where more than two pieces are required for wall height. Stagger the end joints to form a pleasing pattern on the wall, and avoid positioning two or more end joints near each other. Vary the widths to enhance the random effect. For best contrast, use narrow planks adjacent to wider ones. Figure 6-22 illustrates three patterns of panel arrangements. The channel rustic patterns provide strong vertical accents with bold shadow lines, while the bevel-edged tongue-and-groove and shiplap patterns offer the more subtle V-groove effect. Where no accent line at all is desired, square-edged adaptations are used to create tight, flush joints. Most patterns may be installed either vertically or horizontally; the choice is yours.

Installation Techniques. For most vertical applications over plaster or similar walls, 1" by 2" or 2" by 3" furring strips installed (glued) horizontally at 16" or 24" centers are recommended. Where the wall is uneven or wavy, wooden wedges or shims should be used behind the furring strips to bring them into an even line. Clean the furring strip with a wire brush. (The same technique can be used for newly studded walls.) Using a caulking gun and following the adhesive manufacturer's recommendations, apply a continuous bead 1/8" to 3/16" wide of paneling adhesive to the furring strips. Starting at one corner, the first piece of paneling should be plumbed vertically with a level. This may necessitate trimming the corner edge if the wall corner is not plumb. After the fit is assured, place the panel in position and press it firmly to even out the adhesive and assure a tight bond. To hold it in place, tack the panel at the top.

After the first panel is securely fastened, the remaining boards can be installed. Each tongue should be fitted tightly into its groove by being rapped smartly with a hammer insulated by a tongued scrap of wood. Warped lengths should be discarded. If possible, use only full-length pieces that extend from floor to ceiling, except where the wall is masked by bookshelves or other built-ins. Corners must be solid, which will usually require ripping at least one board for its full length to take off the tongue (Fig 6-23). Fit the solid paneling as closely as possible around the untrimmed door opening thickness. (The door, window, and moldings should be removed before starting the job.) Fur out window and door frames to equal the thickness of the furring strips plus the thickness of the paneling. To give the job a finished look, use molding around windows and doors, along the floor and ceiling and wherever else applicable.

Fig. 6-22: Patterns of board arrangement: (left to right) channel rustic pattern; shiplap pattern; and tongue-and-groove pattern.

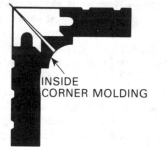

Fig. 6-23: Finished inside corner with standard inside molding (A) or using two butt planks (B).

Solid wood paneling can be applied horizontally or vertically on an existing wall that is reasonably sound and true. The panel adhesive may be applied with a caulking gun or by putty knife. In the latter application, place daubs of adhesive on the back of the panel. These should be about the size of a golfball, at least 1/2" thick and spaced 18" apart near both edges of the panel. Place the panel in position and press it firmly to even out the adhesive and assure a tight bond. Succeeding panels are treated the same: placed close to the previous panel, pressed into place, and then slid tightly against it.

If the old wall surface is masonry, or if it is in poor condition or not true, furring strips should be installed as for vertical solid paneling except that strips should run vertically rather than horizontally. Shim the strips to obtain a true gluing surface. Inside corners are formed by butting the panel units flush with the other walls. If random widths are employed, boards on adjacent walls should be well matched and accurately aligned. Apply the panel adhesive to the furring and press the panels into place. When gluing the boards, be sure to install them so that the tongue edge is out. At the top of the wall, be sure to leave an expansion space of about 1/4". A cove or crown molding will cover it.

Exterior Paneling and Siding. To install exterior sheeting (either plywood or hardboard) or prefinished siding to wood studs, a typical procedure is as follows:

1. Apply a bead 3/16" in diameter of plywood panel adhesive to the framing on all central supports, or where the panel covers the entire member. When two panels join on a single member, apply a double bead on each framing edge so that each panel has continuous contact with the adhesive.

2. Nail at each corner and on 24" intervals around the perimeter of the panel *only*. No nailing is usually required on the central supports. The normal bow of the panel will provide pressure for the bond on the center supports, but extra assurance will be obtained by applying uniform external pressure or blows with a well padded hammer or mallet at the contact lines to flatten the adhesive bead.

3. Nailing should be completed within 30 minutes of the start of the application of the adhesive bead.

CERAMIC TILE WALLS

There are many types of ceramic tiles, including glazed wall tiles, ceramic mosaics, quarry tiles, and specialty tiles, and they are available in a wide range of designs and colors. Actually, tiles range in size from small 1" square mosaics to impressive 12" squares. They are available in high or low relief designs with colorful glazes or multicolored patterns. There are also handsome contoured tiles. Along with hexagons, octagons, and rectangles, there are curvilinear shapes inspired by historic Moorish designs, houses in Normandy, and villas in Florence. Many of these are quarry tiles which are now offered in a large range of natural colors. Because of the many practical and decorative virtues uniquely its own, tile has a place in any room in your home (Fig. 6-24).

In the past, tile could be installed only by highly skilled craftsmen since the only setting material available was cement/mortar, which is difficult to work with and requires the experience of the professional. Today, as a result of the development of new adhesives, it is possible to install your own tile.

To estimate the number of tiles required, add the length and width of a room, multiply by 2, then multiply this answer by the height of the wall to cover. This will give you the square footage of the area to be tiled. For example, for an 8 by 12 room, add 8 and 12 (20) and multiply by 2 (40). If the wall is to be tiled up to 5 feet in height, multiply 40 by 5. This gives a total of 200 square feet to be tiled. If you plan to use common 4-1/4" square tiles, then eight will cover a square foot of surface; thus 200 times 8, or 1600, tiles are needed. The factor of 8 already takes

Fig. 6-24: Ceramic tile may be used for both walls and floors (Courtesy of American Olean Tile Company).

into account waste and allowances for doors and windows. It is always a good idea to get a little more than you actually require in case of accidental breakage and for future replacement should a tile become cracked. You must remember that tiles, even the same color and made by the same manufacturer, are made according to dye lot and shade. Make sure all of your tiles are the same shade and dye lot number.

There are many special trim shapes in glazed ceramics, allowing you to make both outside and inside corners, cove base, bullnose, tub enclosures, tile wainscot borders, and even countertop and windowsill edge pieces. Take a rough sketch of your room along with you to your dealer so he can help you specify these special shapes. These are generally sold by the lineal foot or by the piece.

Before installing the tile, make sure that surfaces are plumb, sound, and free of old materials that might later loosen and cause the new tiles to fall out. That is, loose plaster must be removed and the holes refilled with sound material. If you are tiling over old tile, any loose tiles must be reglued or removed and their holes filled with spackle. Decorative wood wall paneling is often installed with a minimum of fasteners because an unblemished surface is desirable. Be sure that these panels are nailed down securely before applying tile over them. The bond between the tile and the wall is only as strong as the surface of the wall. For this reason, the surface must be clean and prepared for adhesive. Be sure to remove all traces of wax, oils, loose paints, and anything else that would get between the adhesive and the wall. You can apply tile over a sound coat of paint as long as it is clean and you have roughed up the surface with sandpaper. You cannot, however, successfully apply tile over wallpaper. It may hold for a little while, but you run the risk of having a whole wall collapse—in one magnificent sheet perhaps—at some point.

Wet areas, such as shower stalls, tub enclosures, laundry floors, or other surfaces that come in constant contact with water, pose some special problems. Water will not penetrate the tiles themselves, but it will penetrate the grout in very small but accumulative amounts and pass into the surface behind the tile. If the backing surface is affected by moisture, the bond between the adhesive and the surface will begin to fail, eventually freeing the tiles. The grout will also begin to fail, opening small cracks that allow more and more moisture in. Surprisingly enough, exterior-grade, marine plywood, and cement-asbestos board are not appropriate for wet areas. Although these materials are not damaged by moisture, they do swell and "move" when wet, physically disturbing the bond between the wall and tile. Conventional wallboard or plaster simply fail in the presence of moisture.

If an existing wall composed of any of these materials is to form a wet wall in your renovation, it must be covered by a surface which will provide a durable support. Water-resistant gypsum wallboard is a suitable surface for wet wall areas, and, of course, previously tiled surfaces, properly prepared, are ideal. If you intend to install tile yourself in wet areas, your local supplier can guide and counsel you in the appropriate methods and techniques.

A tile installation involves three basic steps: (1) applying the adhesive properly; (2) applying the tile; and (3) filling in the spaces between the tiles with a material called grout. Use a latex-based wall type ceramic adhesive.

Most adhesives are applied to the wall with a trowel with a notched edge (Fig. 6-25). It leaves a deposit of adhesive that looks very much like a plowed field, and because of the carefully controlled size of the notches, it leaves just the right amount of adhesive. If you use the trowel carefully, it will be hard to get too much or too little adhesive on the mounting surface. Be sure not to leave any bare

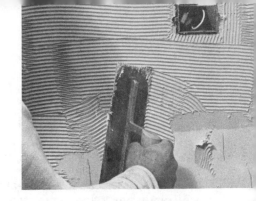

Fig. 6-25: Applying an adhesive with a notched trowel.

spots. For easy working conditions, apply the adhesive to a surface of about 10 to 15 square feet at a time.

The tiles themselves are designed to eliminate some of the careful measuring and precise positioning that an attractive tile job calls for. The smaller mosaic tiles are furnished on a foot-square sheet of paper or scrim, which holds each tile in the right position. You need only locate each sheet on the wall or floor relative to its neighboring sheet. The larger glazed tiles are applied individually. Almost all commercial ceramic tiles have built-in spacing tabs on each edge, which are later hidden by the grout. Handmade porcelain and ceramic tiles, however, do not have these handy spacers. To apply these tiles, simply insert two 6d nails between tiles to give them uniform joints. Leave the nails in until the adhesive hardens and the tiles are firmly in place.

In tiling the wall in any room but the bathroom, begin your tiling at the floor. Draw a center vertical line on the longest wall. Since the wall is almost certainly not exactly the width of a certain number of uncut tiles, the last tiles on each end will have to be cut to fit. Measure from the center line to one end of the wall. If the last partial tile is more than one half the width of a whole tile, begin your installation at the center line with the line at a tile joint. If the last tile is less than half a tile, begin at the center by placing a tile directly over the center line. Spacing this way will prevent an unsightly small fraction of tile at each corner. Work in areas 3' or 4' square so you will not have exposed adhesive on the wall for any length of time. Finish each wall to the ceiling before beginning the next.

When setting the tiles in the adhesive (Fig. 6-26), press them firmly in place, using a slight turning motion. Fill the bottom row first. Leave one tile off the end of each additional horizontal row until you have reached the top of the "staircase." Check to see if both vertical and horizontal lines are straight with a level. If not, the number of tiles needing adjustment will be fewer than if the same number of full rows had been set. If they are level, though, keep adding tiles in the same "staircase" manner, frequently checking with the level. A little extra care spent on the junctions where the tiles come together will greatly enhance the finished work.

Fig. 6-26: Applying the tile in the adhesive.   Fig. 6-27: One method of cutting a tile.

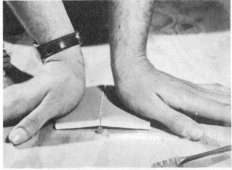

To cut tiles (Fig. 6-27), first mark the glazed surface with a pencil or sharp crayon. If there are ridges on the back of the piece, mark parallel to, not across them. Using a straightedge, scratch along the indicated cut with a steel wheel glass cutter. Now, place a finishing nail on the tile, face up, keeping it in line with the cut. Then, exert pressure on both sides of the tile at the same time to make the break. Smooth any rough edges with a Carborundum stone. Some home handymen quickly develop a knack of giving one edge of the marked tile a sharp blow, which will cause it to break at the indicated place. Professional tile cutters are sometimes loaned or rented by tile dealers. This device is simply a big glass cutter in a frame to hold the tile securely. Simply draw the cutter across the tile, press the handle to make a clean break, and the tile is ready to install.

The shaped cuts, to fit the tile around plumbing, light fixtures, and such, can be made with a glass cutter, then nibbled away with a tile nipper. Take small bites, not large ones. By far the easiest method is the use of a "rod saw," a small abrasive rod that mounts in a hacksaw frame. Use it as you would a coping saw with plywood.

When tiling a bath you begin a bit differently. Find the lowest corner of the tub in its recess. From that corner, measure 1/4" plus the width of one tile up the wall, then mark the place (Fig. 6-28). With a level, draw a horizontal line from this point all around the room. This line will be your reference line for the whole job, and you will begin by tiling up from it. While most adhesives will hold the unsupported tile firmly enough, you may feel a little safer if you tack a furring strip at the line to set the tiles on. When the wall above the line is complete, remove the furring strip, then add the course of tiles below the line (remember the 1/4" plus the width of one tile you skipped). As you work from the low corner of the tub toward the high corner (the difference probably will not be more than a 1/4" or so), you must maintain that 1/4" space by trimming small amounts off the bottom of the tiles with a tile nipper. That space is kept to insure adequate room for sealing between the tub and the wall (Fig. 6-29). When the tub enclosure is complete, go on to the other walls, but start at your reference line so the horizontal lines throughout the room will be uniform. In a bath installed this

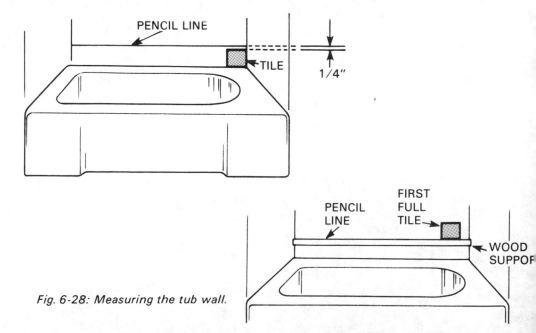

Fig. 6-28: Measuring the tub wall.

way, do not expect the tiles at the baseboard or floor to come out even. You will almost certainly have to trim them to fit.

If you use a ceramic cove base, it is best not to attempt to cut it. Instead, install it in whole pieces before you run the field tile down to the base area. Cut only the ends where it needs to be fitted into a corner, up against the tub, or against another molding area. After installing the cove base, continue to install the field tile down to it. Then, cut the field tile course immediately adjacent to the cove base to fit, with the cut edge adjacent to it. This will give you a much neater installation and will avoid the cracking of the more expensive cove base pieces during cutting.

As soon as all the tiles are in place, remove any spots of adhesive with thinner or solvent. Lighter fluid sometimes works well. Let the job set for a day and it is ready for grouting, or filling the spaces between tiles. Always wear rubber or plastic gloves. Most grouts are caustic and could cause problems for your hands, especially around the fingernails. Latex grouts are generally considered to be superior to regular grouts for floors because they will "give" a little and are not as likely to crack when they dry. Many of the modern grouts come premixed and are available in a wide range of colors.

Fig. 6-29: Applying caulking around a tub.

One of the easiest ways to apply grout to the tile joints is with a sponge, sponge float, or small squeegee (Fig. 6-30A). Force the grout into the joints by stroking diagonally. This diagonal stroke prevents the grout from being "dragged" out of the joint as you proceed. Make sure all joints are completely filled and any air pockets have been eliminated. Grout only about 25 or 30 square feet at a time. A sponge doubled in half works well to grout corners and other areas where the squeegee cannot work. A few minutes after the grout has been applied, a thin film will appear over the entire tiled surface. Wipe it off lightly with a damp cloth (Fig. 6-30B), frequently rinsing the cloth out in clean water. Repeat this process until all the joints have been grouted. Then, to smooth and harden the joints, strike them with a toothbrush handle or the head of a medium-sized nail. Although the nail does the best job, especially on grout that has already started to harden, care must be taken not to damage the remaining grout or to scratch the tile surface. Permit the tile grout to dry completely, usually about 48 hours, before using.

Marble Tiles. Natural marble and cultured tiles (usually 6" by 6" by 1/4") are available which can be used for accent walls in the kitchen, bath, dining, or living room. The natural beauty of marble can turn an ordinary fireplace into the focal

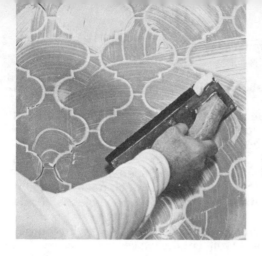

Fig. 6-30: Completing a ceramic tile wall.

point of your den or living room. Marble tiles are installed on the wall in basically the same manner as ceramic tile. That is, after the wall is checked for good adhesion, a wall ceramic tile adhesive is applied about 1/8" thick over as much an area as can be easily tiled in 20 minutes. Then, press the tile into place with a slightly twisting motion; do not slide the tiles as the adhesive will squeeze through the joints. Seat the tiles and level the face area by gently tapping a 2" by 4" with a plastic-faced hammer. Remove any excess adhesive with razor blades; do not clean the tile surface with an adhesive solvent. The grouting of the tiles is accomplished in the same way as for walls of ceramic tiles.

Countertops. A ceramic tile counter will not burn, blister, scorch, or crack. When rubbed over lightly with a damp cloth, a ceramic countertop is as fresh as new (Fig. 6-31). Tile is also just great for vanity tops.

The countertop of ceramic tile is constructed in the same manner as for plastic laminates (page 77). Lay out the edge tiles with full pieces in the corners; adjust the center widths. With the layout in mind, remove the tiles from the countertop and apply a wall ceramic tile adhesive to it. Spread the adhesive evenly with a notched trowel. Install the tile along the edge with a 1/4" overhang so it tops off the edges of the other tiles. Use nippers or a rod saw to shape tile to irregular contours, such as sink corners. Make straight cuts for the opening edge with a hand cutter. Install additional tile; cut to meet the edge wherever it is necessary. Also install the edge strip tiles.

Tiles on the backsplash should exceed its height by 1/4" to allow for the edge tiles. Backsplash strips should be butted against backsplash tiles to form a square corner.

Allow the tiles to dry overnight, then mix and apply the grout according to the manufacturer's directions.

Fig. 6-31: Typical ceramic tile countertops.

BRICK VENEER AND SIMULATED BRICK WALLS

Real stone and brick veneer and the simulated types are installed in basically the same manner. Simulated stone and brick are made of various plastic materials: styrene, urethane, and rigid vinyl are the most common. Some are fire resistant and may be used as fascias for fireplaces. All imitation bricks and stones are highly durable and come in a wide variety of colors and styles (Fig. 6-32). Real stone and brick veneers are generally 3/4" thick and come in several patterns (Fig. 6-33).

When applying either real or simulated materials, be sure to follow the maker's instructions exactly. In all cases, the material should be applied only to a surface that is dry and clean. Remove any loose pieces of wallpaper or paint, and level any bumps or hollows with patching plaster or by sanding so the simulated material will have a flat surface to which it can adhere. If the wall surface is paint, it is usually wise to roughen the surface with a medium-grit sandpaper, scratching through the paint to the subsurface material. On the other hand, if the surface is new plaster or plasterboard, it should be given a prime coat of paint. If the installation is to extend to the ceiling, use the top of the wall as a starting reference line. From the top of the wall, draw a vertical line down the wall on each side of the area to be covered. Use either a carpenter's level or a plumb line to make sure the lines are vertical. If the installation is to run the

Fig. 6-32: Typical simulated brick and stone walls.

entire width of the wall, the room corners can serve as the vertical lines. Incidentally, most manufacturers of imitation brick and stone supply L-shaped pieces for inside and outside corners; they give the finished "masonry" work a more realistic appearance.

Starting along the top line, spread the tileboard or wall ceramic tile adhesive with a 3/16" toothed trowel over a 3' to 4' square area. Then, firmly press the brick or stone into the adhesive with a side-to-side motion, allowing 3/8" to 1/2" mortar joint space. With some installation systems, a caulking gun is used to fill the joints with a special mortar. Set and clean the mortar joints with a dowel as soon as the mortar starts to harden.

When installing stone or brick veneer, it is a good idea to slightly dampen the material with water before pressing it into the adhesive. To cut real materials, score the back with a hacksaw. That is, place it face down on a flat surface, score the back, and snap it over a nail, as when cutting ceramic tile. Simulated material can usually be cut with a sharp utility knife and a metal straightedge.

Fig. 6-33: Typical real brick and stone walls.

INSTALLATION OF CEILING TILE

In many homes, especially older ones, the existing plaster or drywall ceiling is still in sound, level condition, but it is old and unattractive looking. If that description reminds you of a situation which exists in your home, then install new decorative ceiling tile directly to your present ceiling.

The key to installing ceiling tile with adhesive is to be sure that the present ceiling is structurally sound, dry, and free from dust, dirt, oil, or grease. Any loose wallpaper or flaking paint is not critical; it can be removed as long as the surface above it is sound. New plaster, concrete, or other similar surfaces must be fully cured prior to applying the adhesive. The material you use may be acoustical, that is, containing small pockets that absorb sound, or it may be plain. They are available in sizes from 12" by 12" tiles all the way up to large 4' by 10' panels. However, the most popular size is the 12" by 12" (Fig. 6-34).

Fig. 6-34: Typical ceiling tile installations.

Planning the Job

Calculating your ceiling tile needs is simple because the tiles are 12" by 12" or one square foot in size. Just multiply the length of your room by its width to determine the number of tiles needed and add one extra tile or 12" to each dimension. See example. This (plus one tile) allows for cuts, any required fitting, and extra tile.

For example:

Room length: 12' + 1' = 13'
Room width: 10' + 1' = 11'
(13' X 11' = 143)

Total tiles needed: 143

If your room does not measure an exact number of feet (Fig. 6-35), use the next highest numbers as follows:

	Actual Dimensions	Multiply
Room length:	12'8"	13' + 1' = 14'
Room width:	10'6"	11' + 1' = 12'
		(14' X 12' = 168')
Total Number of tiles:		168

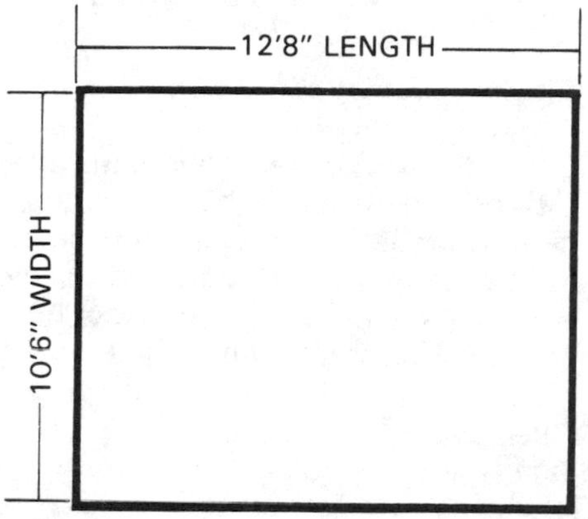

Fig. 6-35: How to measure a room for ceiling tiles.

Add any tiles required for special alcoves or vertical applications over boxed ductwork. Subtract tiles eliminated by skylights, columns, large recessed lighting fixtures, and attic doors. If your room is irregular in shape, plan to arrange your tiles for the best appearance in the largest ceiling area. If you plan to install a recessed lighting fixture between joists, it should be done before installing the tiles. The 12" by 12" ceiling tiles can be applied in a basic square pattern, diamond pattern, or offset to form an ashlar pattern (Fig. 6-36). To figure the total cost, just multiply the number of tiles by their single unit cost to arrive at the total.

Tile Installation. The ceiling tile and adhesive should be conditioned to the room temperature and humidity before application. To accomplish this, open the cartons in the room where the tile will be installed and allow them—along with the adhesive—to stand for at least 24 hours.

Follow all the manufacturer's directions and avoid any tendency to take "short-cut" methods, the most common cause of problems in installing ceiling tile of any make or manufacture. For instance, the most frequently ignored steps in the entire procedure are those involved in determining proper border tile widths and establishing the chalk guidelines essential to a completely successful ceiling

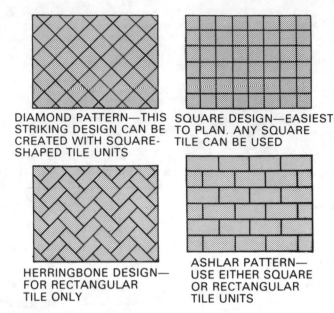

DIAMOND PATTERN—THIS
STRIKING DESIGN CAN BE
CREATED WITH SQUARE-
SHAPED TILE UNITS

SQUARE DESIGN—EASIEST
TO PLAN. ANY SQUARE
TILE CAN BE USED

HERRINGBONE DESIGN—
FOR RECTANGULAR
TILE ONLY

ASHLAR PATTERN—
USE EITHER SQUARE
OR RECTANGULAR
TILE UNITS

Fig. 6-36: Ceiling tile patterns.

construction. Spend the few minutes necessary to complete the simple calculations and establish the guidelines; then enjoy the pleasure of putting up a new
ceiling properly and without problems in the very minimum amount of time.

To determine border tile widths, measure the short wall of the room first. If the
wall does not measure an exact number of feet, add 12" to the odd inches left
over and complete the simple calculations as shown in this example:

Short Wall:	10' 6"
Add:	12"
	————
	18
	————
Divide by:	2 = 9" width for the long walls border tile

In the same manner, border tile width for the short walls is figured as follows;
measure long wall:

Long Wall:	12' 8"
Add:	12"
	————
	20
	————
Divide by:	2 = 10" width for short walls border tile

When using an adhesive to cement tiles, start the work in the center of the
room. To determine the exact center of the room, measure along the sides to
locate the center lines, and then snap a chalk line between the midpoints of each
wall. Checking your calculations against the paper layout, determine the border
widths. Should you be unable to achieve the proper arrangement of tiles (a poor

border layout), snap secondary guidelines half a tile width to one side of the center lines. The first tile will then be set with its corner at the intersection of the guiding chalk lines. By starting the installation in this way, you can work the four equal quarters of the room individually. Of course, if the room is square, the borders will be even. However, when the secondary chalk lines are snapped as just described, the room will not be divided in equal quarters, but the borders will be equal for each pair of edges.

When installing an ashlar tile arrangement, snap the center lines as just mentioned. Then snap another line the short way of the room, spaced half a tile away from the first. Set the first tile as before; but stagger the next one.

To make the work easier, use a plank over two sawhorses. Arrange a box or bench with two piles of the tile, back side up. Sprinkle your hands with cornstarch or talcum powder before handling the tiles. This helps to keep them clean and dry.

For a good bond, apply a golf ball size glob of adhesive with a putty knife or brush on all four corners and in the center of the tile (Fig. 6-37). Keep the adhesive spots about 2-1/2" from the tile edge. Line up the tile for application by placing it lightly against the old ceiling about 1-1/2" from its ultimate desired location. Then, simply slip the tile into position with gradually increasing pressure. It is a good idea to work the tile back and forth about 1/2" as it is pushed into position to flatten out the adhesive glob. Globs should be flattened out to 3/16" thick and about 2-1/2" in diameter. If the surface of the old ceiling should be slightly uneven, increase the thickness of the adhesive to make the finished tile surface level. (Do not pull the tile away from the base to make it level. Rather, remove it and add more adhesive.) When installing interlocking tile, do not slide the tile into the groove of the adjacent tile until it has been pushed into its proper level against the ceiling.

When the first row of tiles has been completed, apply the next one. Continue until the whole ceiling has been tiled. To fit tiles around posts, cut them in two and remove a semicircular portion from each half with a utility knife or fine-toothed saw. Then, fit the two pieces around the post and fasten them up. To fit tiles around ceiling fixtures that are located where four tiles come together, simply notch each tile in the proper corner and fit them carefully in place. When installing the border tiles, allow about 1/4" of space between the tile and the wall for expansion.

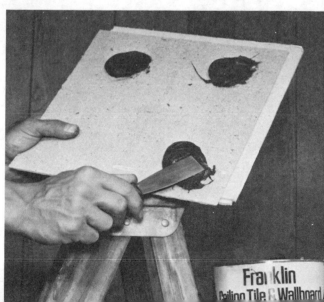

Fig. 6-37: Applying adhesive on ceiling tile.

In any tile installation, when the tiles are in place, you can finish off the room nicely with one of a number of attractive cove or crown moldings. Use 3d to 6d finishing nails and/or construction adhesive to apply such molding. When using nails, be sure to drive them into wall studs whenever possible in preference to nailing through the tile into the joists.

Ceiling installations present one critical problem for adhesive applications— ceiling tile has a dead load weight pulling directly downward. It is therefore absolutely necessary that an adhesive with high cohesive strength be used. Most good quality mastics and construction adhesives do not have this high cohesive strength. It is usually found only in the resin based adhesives that are designed specifically for ceiling application. Therefore, unless you are good at juggling, do not try to install ceiling tile with any adhesives other than those designed for this type of application.

Ceiling Insulation

In many older homes, or those with cathedral ceilings or exposed wood roof decks, the only practical way of insulating the ceiling from inside is to apply plastic foam to the present ceiling, and then bond a finished ceiling to the insulation. Frequently in new construction, foam insulation is sandwiched between two thicknesses of 1/2" gypsum board. Before applying the foam, be sure the surface that you will be bonding to is structurally sound, clean, dry, and flat, and is free of grease, loose paint, and wallpaper. Remove any ceiling molding. Also, much heat is lost due to cold air blowing in through cracks. Take this opportunity to caulk any cracks where the ceiling and walls meet; a good acrylic latex caulk is ideal for this purpose.

To install the insulation, apply a foam adhesive to the foam, following the instructions on the cartridge. Adhere the foam insulation to the ceiling, running it whichever direction is more convenient for you. Press the foam firmly and uniformly across every square foot to assure an intimate bond. Drive three or four roofing nails into each foam board to hold the foam in close contact with the ceiling while the mastic is curing. Once the foam is in place, the ceiling tile of gypsum board may be installed as previously described.

Putting Down Floors with Adhesives 7

Fine flooring has been a symbol of luxurious living since the days when noblemen and kings walked on marble and rich parquetry. Today, floors are more important than ever in the decorating scheme—but wealth is no longer a prerequisite for enjoying luxury underfoot.

The choices of floor coverings are categorized into the following groups: wood, resilient surfaces, rigid (ceramic tile, slate, and brick), and carpeting. All of these floor coverings can be easily installed with the proper adhesives.

TYPES OF ADHESIVES

For many years, the only adhesives that could be used for the installation of floor products were portland cement and black asphalt emulsion mastics. As mentioned earlier in the book, a "mud" of portland cement was used to hold ceramic tiles. The black asphalts were at one time widely used, but because of the oozing between the tiles, they have been generally replaced by the emulsion latex type adhesives. Both, however, are still in limited use today. For example, asphalt mastics—either of cut-back or emulsion types—are one of the few adhesives that can be used with polyethylene film—one of the most difficult substances to bond. Except for a few limited uses discussed later in this chapter, asphalt mastics have given way to the new synthetic types (Fig. 7-1) that follow.

Fig. 7-1: Some of the more common floor adhesives.

Subfloor Adhesives. These adhesives were specifically developed for bonding plywood directly to floor joists, but have now developed into a general purpose use category in many cases. Most of these multipurpose formulations achieve exceptional bonds to woods, concrete, gypsum wallboard, insulation board, and similar construction surfaces. Their high degree of initial tack allows light pieces to be bonded with only momentary pressure. The allowable open time is regulated by the temperature, humidity, and air circulation. The porosity of the substrate or adherend will also have an effect on the drying rate, depending upon the ability of these surfaces to absorb solvent from the system. Irregularities and voids in substrates, such as framing, are readily bridged with subfloor adhesive, providing a more solid back-up surface.

One of the most common methods for applying subfloor mastics is by extrusion, either in cartridge form or specialized extrusion equipment in factory use. Where cartridges are used, so many variables are present that it is difficult to actually pinpoint the coverage. Where the nozzle tip is cut, speed of application, whether the bead is continuous or broken, etc., all have a great effect on coverage. However, the data in the following table may be helpful in estimating volume requirements for various bead lengths and sizes.

Volume Extruded	Length of Bead Extruded With Various Bead Diameters		
	1/8 in.	1/4 in.	3/8 in.
1 U.S. Gallon (128 fluid ounces)	1,569 ft.	392 ft.	174 ft.
Quart size cartridge (29 fluid ounces)	355 ft.	89 ft.	39 ft.

Since the above data is calculated on the basis of cubic measurements as related to fluid measurements, these figures will provide a reasonably accurate measure of volume required for most any mastic adhesive.

When selecting a subfloor adhesive, be sure that it meets all of the requirements of the American Plywood Association (APA) Specification AFG-01 and HUD/FHA Use of Materials Bulletin #60. Commercial cleanup solvents available that would be recommended are mineral spirits and, on small jobs, lighter fluid may be helpful. **Caution:** Never forget that many solvents are extremely flammable and some are even toxic. Be extremely cautious when using them. Never use gasoline or similar type oily solvents, as they may leave an oily film on the surface that will not permit a satisfactory bond to develop. Where the adhesive has dried, it is best to scrape all excess off with a putty knife. In some cases, soaking may be necessary for several hours to completely loosen the dried adhesive. **Note:** Always make certain that the solvent used will not mar or attack the surfaces to which it is being applied by testing an out-of-the-way area.

Floor Tile Adhesives. Most multipurpose floor tile adhesives are water resistant both indoors and out after completely drying. They are suitable for vinyl tile, vinyl asbestos tile, vinyl roll goods, rubber tile, asphalt tile, and carpeting. In addition, they may be used on floor surfaces such as plywood, particleboard, and smooth, dry concrete either below, on, or above grade, providing that there is no moisture or hydrostatic pressure in the floor. Most of these multipurpose adhesives permit adjustment of the flooring material after they are applied.

So-called "fast dry" floor tile adhesives dry quite rapidly into a "semi-pressure sensitive" state, allowing ample time for laying the tile. The dry film remains transparent so that chalk marks or other guide lines can easily be seen. It may be used on most floor surfaces such as plywood, particleboard, hardboard, and smooth, dry concrete (above, on, or below grade), again providing there is no moisture or hydrostatic pressure in the floor.

Most floor tile adhesives are easy to clean-up. Wet adhesive on the face of the tile may be removed with warm soapy water. If the adhesive has set up, remove it from the tile with a soapy steel wool pad. Tools may be cleaned with a clean-up solvent.

Cove Base Adhesives. These resin base formulations are designed specifically for the easy installation of rubber and vinyl cove base materials. They develop excellent adhesion on wall surfaces such as plaster, gypsum wallboard, wall papers, plywood and hardboard paneling, ceramic tile, tileboard, etc.

Carpet Adhesives. These adhesives were developed specifically for installation of residential and commercial carpeting with impregnated, laminated foam rubber backing, or plain jute backing. (Most general-purpose carpet adhesives are not recommended for vinyl foam backed carpeting; special adhesives are available for this type of carpeting.) Also, certain types of general carpet adhesives can be used on popular indoor/outdoor type carpeting for patios, porches, around swimming pools, in kitchens, family rooms, and recreation rooms. It provides a quick bond between carpet and floor surfaces of wood, plywood, particleboard, and concrete; and it can be used above, on, or below grade level. They apply easily with a trowel, spread well, and have no offensive odor.

Wood Floor Mastics. These products are specially designed for the installation of wood parquet, strips, or planks. They may be applied over smooth, dry concrete, plywood, particleboard, or hardboard surfaces. The use of a latex emulsion system for wood flooring is not recommended. Remember, wood does not like water. For this reason, emulsion systems have not been generally acceptable to the flooring industry because of the wood absorbing the water and causing excessive contraction, expansion, and warping.

Ceramic Floor Tile Adhesives. These solvent and latex adhesives are similar to the ceramic adhesives used for walls, except that the floor adhesives generally dry more rapidly than wall types. While you can spread adhesive over a fairly large area of wall before you set the tile, it is better to work in small areas when applying tile to the floor. Floor tile adhesive can be applied by a trowel.

Epoxy cements are catalytic, two-part cements that are used in some ceramic floor tile installations. While these cements are popular with professional tile men on certain applications, because of their peculiar setting characteristics, most do-it-yourselfers would not find them too adaptable for their applications.

In addition to the so-called "floor" adhesives, the general-purpose construction adhesive described in the previous chapter also has flooring applications. As with all adhesives, select the floor mastic that meets your needs—setting time, bonding ability, and water resistance—and follow the directions of the adhesive manufacturer.

THE SUBFLOOR

In most remodeling or refurbishing jobs around the home, the present floor will act as the subfloor for the new flooring material. However, when finishing off an attic, basement, or garage, or when adding a room to your present house, a subfloor must be installed to receive the finished flooring material. In this type of new construction work, adhesives can play a very important part. In fact, since the mid-1960's the American Plywood Association has been carrying on an intensive test program to perfect a subfloor system that would improve floor quality and performance while using less material. The Association's present system is based on recently developed gluing techniques and adhesives that firmly and permanently secure the structural plywood underlayment to wood joists. The glue bond is so strong that floor and joists behave like integral T-beam units. That combination increases stiffness when compared with conventional floor construction. For instance, stiffness of the joists is increased significantly (about 25% with 2 x 8 joists and 5/8" plywood) when a single layer of plywood underlayment is glued to lumber joists. Gluing the tongue and groove joint between panels approximately doubles the increase in stiffness (to about 50% in the case mentioned). Thus, glued floors not only deflect less under traffic, but tests show better resistance to long-term deflection than nailed-only floors. Gluing also helps to eliminate squeaks, bounce, and nail popping.

APA Glued Subfloor System

Basically, the glued floor acts like a one-sided stressed skin panel; plywood adheres solidly to framing members by means of a strong glue bond (Fig. 7-2). Like the stressed skin panel, it is outstanding in its efficient use of materials while reducing size and weight. Perfection of the system was made possible by a breakthrough in the development of adhesives, which permits them to be used effectively for field-gluing, even in below freezing weather, using materials and techniques readily available on the construction site.

Only glues which conform to Performance Specification AFG-01, developed by the APA to assure dependable quality construction, are recommended for use with the APA Glued Floor System. AFG-01 requires adhesives to develop high levels of shear strength under a wide variation of moisture and temperature conditions, to possess gap-filling capability, and durability with respect to moisture and oxygen exposure. Conformance with these requirements should be certified by a qualified independent testing agency. FHA, in Use of Materials UM-60, requires that initial testing of glues recommended for the APA Glued Floor System must be followed up with a continuing field testing program.

For most subfloor work, 3/4" thick, tongue-and-groove plywood is recommended. The APA has rather detailed literature regarding the glued floor system and the joist size, spacing, and span table data. This information is available from any local American Plywood Association office or can be obtained by writing the American Plywood Association, P.O. Box 2277, Tacoma, Washington 98401. It should be noted, however, that while all of the APA recommendations have been properly engineered and tested, it is a good idea to check local building codes to make sure the grades and species of lumber meet local requirements. Keep in mind, however, that no codes will restrict the use of adhesive if conventional joist spans and fastening schedules are used with plywood single floor construction. The addition of any adhesives is in excess of minimum requirements.

Installing the Subfloor. The steps involved in laying a subfloor employing the APA glued system are as follows:

1. Snap a chalk line across the joists, 4' in from the wall (Fig. 7-3A). This will act as a guide for the edge alignment of the plywood subflooring and may be used as a boundary for extruding the adhesive.

2. Wipe any mud, dirt, or free water from the joists before applying the adhesive. Apply the adhesive by extrusion from cartridges (Fig. 7-3B) or automatic extrusion equipment. A bead of approximately 1/4" to 3/8" thickness should be extruded along all floor joists. It is usually recommended, especially for the do-it-

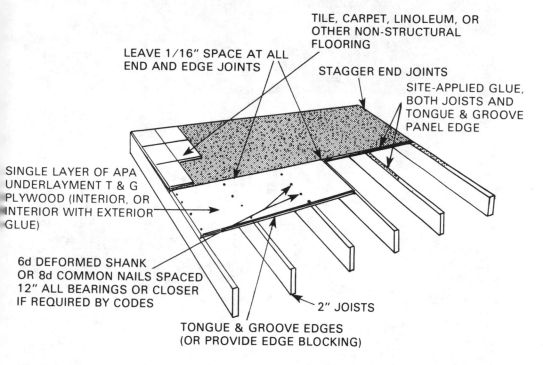

TILE, CARPET, LINOLEUM, OR
OTHER NON-STRUCTURAL
FLOORING

LEAVE 1/16" SPACE AT ALL
END AND EDGE JOINTS

STAGGER END JOINTS

SITE-APPLIED GLUE,
BOTH JOISTS AND
TONGUE & GROOVE
PANEL EDGE

SINGLE LAYER OF APA
UNDERLAYMENT T & G
PLYWOOD (INTERIOR, OR
INTERIOR WITH EXTERIOR
GLUE)

6d DEFORMED SHANK
OR 8d COMMON NAILS SPACED
12" ALL BEARINGS OR CLOSER
IF REQUIRED BY CODES

2" JOISTS

TONGUE & GROOVE EDGES
(OR PROVIDE EDGE BLOCKING)

Fig. 7-2: APA glued floor system (Courtesy of American Plywood Association).

yourselfer, that only sufficient adhesive be applied at one time to install one or two 4' by 8' sheets of plywood. Of course, always follow the specific recommendations of the adhesive manufacturer.

3. Lay the first panel with the tongue side to the wall, and nail it in place with 6d deformed shank or 8d common nails spaced 12" apart. Such panel placement protects the tongue of the next panel from damage while being tapped in place (Fig. 7-3C). Complete all nailing of each panel before proceeding with the adjacent panel.

4. Apply a bead of glue (about 1/4" diameter) to the sill and framing members. On wide areas, apply the adhesive in a serpentine pattern. Also, apply two beads of adhesive where the panel ends butt, to be sure each end is glued (Fig. 7-3D).

5. After the first row of panels is in place, extrude a bead of adhesive into the grooved channels (Fig. 7-3E) of one or two panels at a time before laying the next row. The adhesive bead should be continuous on all joists. Avoid excessive squeeze-out on the tongue and groove by applying a thinner bead than on the joist.

6. Use a block to protect the groove edge when tapping the panels into place (Fig. 7-3F). Stagger the end joints of the panels in each succeeding row, leaving a 1/16" space between all end and side joints (Fig. 7-3G). Most carpenters space panels "by eye" to assure the 1/16" spacing. Remove any adhesive squeeze-out.

7. Complete the nailing of each panel before the adhesive sets (Fig. 7-3H). Space the nails 12" on center along the joists. Set the nails 1/16". The finished deck can be walked on and carries construction loads without damaging the glue bond providing **all** nailing of each panel is completed as you proceed.

Fig. 7-3: Steps in installing a subfloor by APA glued floor system.

Subfloor on Metal Beams

While the average do-it-yourselfer would never have call for this application, it might be interesting to note that the construction industry is now using construction or metal framing adhesive to secure plywood subfloors to steel and aluminum beams, as well as other metal framing parts. This system uses a combination of neoprene structural adhesives and self-drilling screws which adds rigidity and unitized strength to the subfloor and the overall structure. As a result, "callbacks" or requests for repairs because of squeaky or warped floors or popped nails have been virtually eliminated, acoustical qualities are enhanced, and owner satisfaction is widely evident.

Subfloor Over Concrete

Where an insulated concrete floor is desired, it is possible to install foam insulation board under a plywood wearing surface. Before applying the insulation board, be sure the floor surface you will be bonding to is structurally sound, clean and dry, and is free of oil and loose paint. Any cracks should be filled with patching cement or a suitable concrete latex underlayment several days prior to installation. Much heat is lost due to cold air blowing in through cracks, so take this opportunity to caulk cracks where the floor and walls meet; concrete binder works fine for this purpose. If your concrete floor has a drain, you will probably not want to cover the drain, unless you can determine that your local code allows it to be covered.

Determine which direction the plywood should run and plan to run the furring strips at right angles to the plywood. The foam insulation board can run whichever direction is easiest for you. To install the subfloor, proceed as follows:

1. Cut 2" by 2" wood sleepers or furring strips. These may be cut by merely ripping 2 by 4's in half. Using a chalk line, strike a line on 16" centers beginning at the wall that will run perpendicular to the direction of your plywood installation.

2. Apply a serpentine bead of a good quality construction adhesive at the area of the chalk line and position a 2 by 2 along this line. Continue applying the 2 by 2's on each chalk line spaced 16" on center apart.

3. Next, cut the foam to fit between the sleepers. Using an adhesive specifically recommended for the installation of polystyrene foam, apply beads around the perimeter and in an x pattern to the back of the foam and press into position on the concrete between the sleepers (Fig. 7-4).

4. Refer back to the section on "installing the subfloor" and follow the directions described.

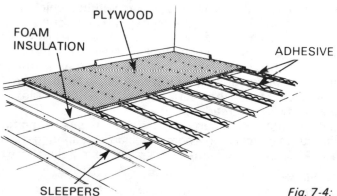

FOAM INSULATION PLYWOOD ADHESIVE

SLEEPERS

Fig. 7-4: Insulated floor over a concrete slab.

5. You are now ready to apply any additional flooring on top of the subfloor, such as carpeting, wood parquet, vinyl, etc.

WOOD FLOORS

Wood floors were probably first used in the mid-11th century and have been a leading flooring material ever since. Today, both hardwoods—oak, birch, beech, and maple—and softwoods—pine and fir—are used for wood flooring.

There are several types of wood floors—solid block parquet, laminated parquet, strip, and plank (Fig. 7-5). Which you choose depends on which appeals to you most. For instance, solid block parquet flooring is available in a number of different styles and designs. They are usually made from three or four short strips of flooring glued together or joined with splines and are most commonly available in a thickness of 25/32". Block sizes are 5", 6-3/4", 7-1/2", 9", and 11-1/4" squares; most popular is the 9" size. Some manufacturers also produce rectangular blocks in a 3/8" thickness.

Fig. 7-5: Three types of wood flooring: (left to right) block, strip, and plank (Courtesy of Bruce Hardwood Floors).

Strip wood flooring is usually made in widths of 1-1/2", 2", 2-1/4", and 3-1/4", with the 2-1/4" size being the most popular. Most strip floors are composed of pieces of uniform width. The strip thicknesses range from 5/16" to 25/32" and more. Square-edge strip flooring is made in 5/16" by 1-1/2" and 5/16" by 2".

Plank wood flooring is wider than the strip type; they usually range from 3" to 9" in width. They are generally laid in random or mixed widths, which gives a somewhat less formal floor than the strip type. Modern plank flooring is often bored and plugged at the ends to simulate Early American types. In Colonial days, wooden pegs were used to fasten down such planks instead of nails.

All wood floor types can be had prefinished. This means that when the new floor is laid, there is no need for sanding, and therefore, there is no dust or finishing odor. And you get a better finish than you could put on yourself. Most prefinished floorings are already waxed ready for use.

Block, strip, and plank wood flooring can be glued over most existing floors when remodeling or directly to a subfloor in a home addition. By adhering sleepers to a concrete floor (page 125), strip and plank flooring can be installed here too.

Preparation for Wood Floors

Wood flooring can be installed over subfloors or concrete, dressed and matched wood, or plywood. Under certain conditions, old surface floors of wood or resilient tile are equally suitable. Depending on type, subfloors should answer these specifications:

Concrete. The subfloor should be sound, level, dry, smooth, and clean. Remove grease or oil stains. Level the high spots with a terrazo grinder or carborundum stone. Fill the low areas to general subfloor level with a good quality concrete bonder or latex underlayment (page 171). But, before using any wood product on or below grade, first consult the flooring manufacturer's recommendations on installation over concrete.

Plywood or Wood. Wood flooring can be glued to either of the two subfloor systems described earlier in this chapter. New flooring can be installed over old wood floors if their surface is sound, level, and well-nailed. It is a good idea to rough-sand the old wood surface floors to remove varnish, paint, shellac, or wax. If the wood surface is poor, it may be necessary to lay an underlayment as described later in the chapter for resilient flooring.

Resilient Tile. Before laying a wood flooring directly over old asphalt or vinyl asbestos tiles, it is wise to check the adhesive label. Mastics containing solvents of various types have been known to not be compatible with the old tile and could also not be compatible with previous adhesives used for installing the older flooring. You should be very cautious when installing new flooring over old flooring because of this. Follow the manufacturers directions very closely and, if possible, make a small test installation in some out of the way area to check out the compatibility of the system. If the label on the container indicates that this is an accepted practice, and if the tiles are not crumbled, loose, or otherwise in poor condition, then the wood flooring material can be installed directly over the old tile. (Never lay tile over rubber tile.) But, if you have any doubt, or if the tile is in bad condition, apply an underlayment as described on page 138 or remove all the tile down to the subfloor level and sand or scrape the subfloor to remove all traces of the old tile cement.

When installing any of the three types of wood flooring, draw a rough plan of the room, plus the important dimensions (Fig. 7-6). Then, figure the square footage of the room or area, and add 5% to the total for a final quantity figure. Take this figure and plan to your flooring dealer. He will advise you on the amount of flooring and adhesive you will need.

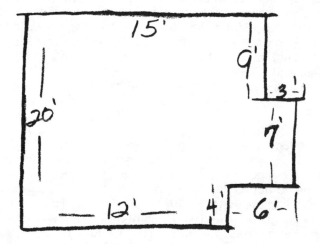

Fig. 7-6: Rough plan of a room.

Wood Parquet (Solid Block or Laminated)

As mentioned earlier, solid wood block parquet or laminated parquet can be installed over concrete slabs, wood, or plywood subfloor, or over existing finished floors in certain instances. They can even be used in the basement if the concrete floor is completely dry and has been properly treated with a suitable vapor barrier system underneath the concrete slab. Never attempt to install any flooring directly over and to polyethylene sheet. Wood blocks also make fine wall and divider decorations on vertical surfaces (Fig. 7-7).

Fig. 7-7: Decorating with wood blocks (Courtesy of Bruce Hardwood floors).

Laying Out Working Lines. When laying out a square pattern, it is necessary to find the starting point as follows:

1. Measure along one wall to find the center point.

2. Locate the same center point at the opposite end of the room or as close to the center as you can get (disregard small alcoves, offsets, and other breaks).

3. Snap a line by rubbing chalk on string, holding the string taut on the two center points. Then, snap the string. This will mark a center line across the room.

4. Find the center point of the other two walls. (In all measuring, disregard bays, alcoves, and offsets.)

5. Before snapping the crossline, be sure it is exactly at right angles to the first center line. This can be determined with a piece of tile or a large carpenter's square.

6. Snap this crossline onto the under floor. Now the main portion of the room (disregarding alcoves, etc.) is divided into quarters. To check right angles, make a 3' by 4' by 5' triangle as shown in Fig. 7-8. If the crosslines are truly at right angles, the three sides of each triangle must measure exactly 3', 4', and 5' as shown. If they do not, swing the string until these measurements are correct.

Before spreading the adhesive, lay a row of loose blocks along the chalk lines, starting at the center point and working out to one side wall and then to the other end wall. This determines the space left for the outside row of blocks (the border). Measure the distance between the wall and the last unit. If the distance is less than 2" or more than 8" (for 9" square blocks), move the center line parallel to

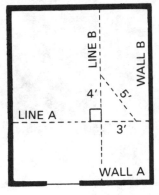
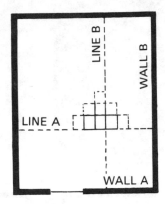

Fig. 7-8: Laying out a square pattern.

and 4-1/2" closer to that wall. This will prevent the peculiar look of blocks that are too small along either wall. If the central point is moved and either line is re-snapped, check both lines again to be sure that they are at perfect right angles to one another.

To lay out a diagonal pattern, proceed as follows:

1. Find the center of the room in the same manner as for a square pattern.

2. From the intersection, measure 4' along each line toward all walls and mark the points established (Points **a, b, c, d,** Fig. 7-9).

3. Using a radius of 4', establish Point **A** by scribing intersecting arcs from Points **a** and **b,** and Point **B** by scribing similar arcs from Points **c** and **d.**

4. Snap a chalk line (Line **AB**) across Points **A** and **B** and extend to nearest wall. This establishes a 45° angle starting line.

Laying Wood Blocks. For the best results, store the flooring and adhesive in a dry area at room temperatures at least 72 hours prior to installation. The mastic should be applied directly from the container without heating. Spread the adhesive in a triangular shape (Fig. 7-10A), being careful not to cover the working line. Use a trowel having square notches approximately 1/4" deep, 1/4" wide, and spaced 1/4" apart (Fig. 7-11). Remember that through continued use, steel spreaders will wear down and result in too small a notch for adequate coverage. When this occurs, the notches should be refiled or a new spreader should be used. Also, the trowel should be cleaned whenever the mastic begins to build up and reduce the notch size. Some floor mastic manufacturers suggest waiting for 30 to 60 minutes after spreading the adhesive before installing the flooring. This time period will improve the tack or green strength and result in less tendency for the flooring to slide and the development of better bonds.

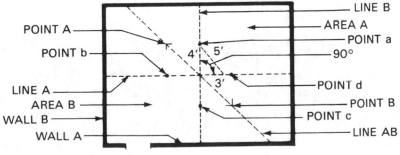

Fig. 7-9: Laying out a diagonal pattern.

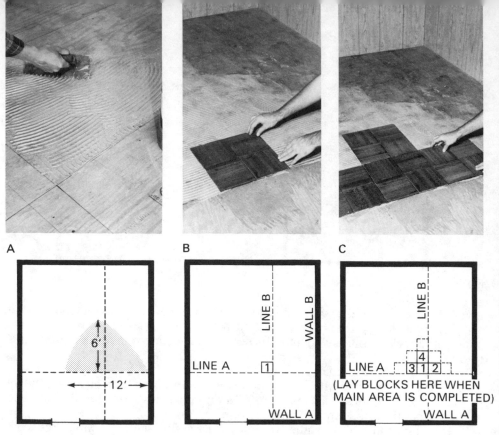

A   B   C

Fig. 7-10: Steps in laying wood block in a square pattern.

When laying the blocks, place tile unit No. 1 on Line A, at a 90° angle point (Fig. 7-10B) along the working line. Then, lay units 2 and 3 on either side of the first unit. Align these units carefully with Line **A** and with each other. Place unit 4 directly over unit 1, as in Fig. 7-10C, establishing a "pyramid" pattern extending from Line **A**. Use the pyramid in the same manner when installing a diagonal pattern. Continue laying around the pyramid in the same manner until the walls are reached. Make sure that the units along Line **A** are positioned exactly on the line to avoid crooked joints. Also, when positioning units, drop them lightly into place with a slight twisting motion and without sliding. Sliding them into place will cause adhesive to pile upon the leading edge, impeding the fit. Always apply the wood floor mastic at the manufacturer's recommended rate of coverage. Too little adhesive may result in "poor bond." Too much may result in "bleeding" between parquet units. Remove a unit occasionally to be sure the adhesive is "transferring" to the back of the wood squares.

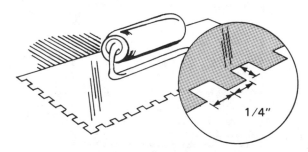

Fig. 7-11: Typical notched trowel used to apply floor mastic.

Fig. 7-12: Measuring and cutting a border block.

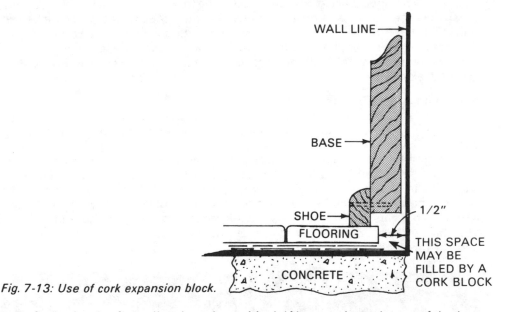

Fig. 7-13: Use of cork expansion block.

To fit the border floor tiles, lay a loose block (A) squarely on the top of the last cemented unit nearest to the border space (Fig. 7-12). Place another full block, 1/2" away from the wall, on top of the middle unit (A). Then, mark the middle block (A) with a pencil along the edge of the top unit (B). The 1/2" space is allowed at the wall for expansion and is covered by a shoe and/or base molding. Frequently, the expansion area is blocked with pieces of cork (Fig. 7-13) or some other soft material.

The wood blocks can be easily cut with a saber (jig) saw. Any obstructions can also be cut with this saw; but, again be sure to leave 1/2" space for expansion. When laying the border units, let them extend as far as possible into the door openings. The parquet squares may be scribed to door jambs or placed under if the jambs are raised up. As the installation proceeds, lay plywood or boards over the installed units to prevent foot traffic from sliding the installed floor units out of place. Methods of laying the blocks for a diagonal pattern are shown in Fig. 7-14.

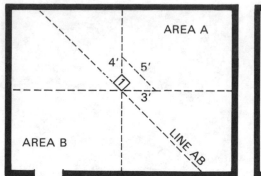
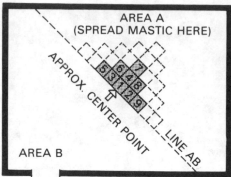

Fig. 7-14: Laying wood blocks in a diagonal pattern.

Most wood block parquet and laminated wood parquet on the market today are completely finished, waxed, and polished at the factory. Do not job-apply surface finish over the factory prefinished surface. Let the adhesive set overnight before replacing the furniture, however. If desired, additional wax may be applied after the installation is completed. Most manufacturers recommend the type of wax and cleaner to use with their floors; follow their advice, as well as the directions on the wax container.

Plank and Strip Flooring

Plank or strip wood flooring can be laid in adhesive in much the same manner as parquet squares. To lay out a room and install random width plank flooring such as shown in Fig. 7-15, proceed as follows:

1. Before beginning actual installation, provide proper layout of flooring by distributing short and long lengths equally over the areas to be floored (planks should be laid at right angles to floor joists), and establish the desired pattern by distributing the face widths. Be certain to use an equal number of rows of each width. Avoid clustering end joints. When installing the type shown in Fig. 7-15A, be sure to cut the first board in each row to avoid unattractive "line-up" of pegs at one wall line.

A   B

Fig. 7-15: Two types of random width plank flooring (Courtesy of Bruce Hardwood Floors).

2. Measure 15-1/2" out from the long starting wall at two distant points (Fig. 7-16A) and mark Points **A** and **B**.

3. Snap a chalk line the length of the room and through Points **A** and **B** (Fig. 7-16B).

4. Start spreading adhesive on the short side of the chalk line and work your way back to the chalk line, leaving the large section as a work area (Fig. 7-16C). For good bond strength, let the adhesive set for about 30 minutes to 1 hour (depending on temperature, humidity, and air circulation) before laying planks; but, again follow the flooring and adhesive manufacturers directions on the label. When the short area is completed, finish the large work area in the same manner, and cut the last row to fit at the wall. Do not spread more adhesive than you can cover in 2 to 3 hours; otherwise, the adhesive may become too dry for good bonding. *Never let the adhesive set for more than 4 hours before laying planks.* **Note:** With most floor adhesives, when you put new adhesive over adhesive that is almost dry, the old adhesive softens up and is usable, so do not scrape it away unless the manufacturer's instructions so state or it has set up for too long a period.

Fig. 7-16: Steps in installing plank flooring.

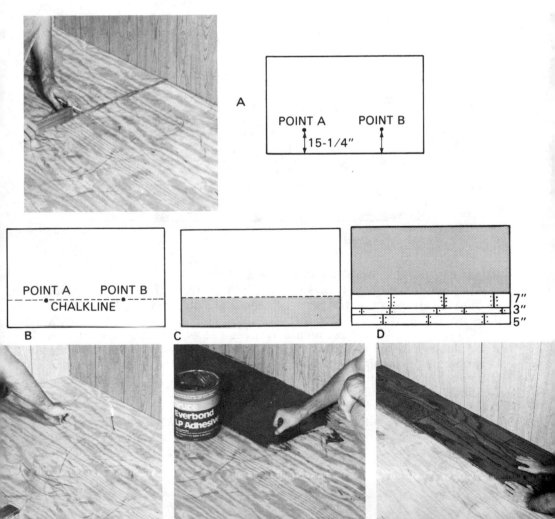

5. First lay the 7" row of planks with groove facing chalk line and work toward the opposite wall. Be sure to keep fingers out of adhesive to avoid excessive cleanup later. Press each board in place and proceed to the next row (Fig. 7-16D). After you put down a section of flooring, roll with a 150-pound roller to be sure all boards are in firm contact with adhesive. Remove adhesive smears with mineral spirits, xylene, toluene, or lighter fluid.

6. Occasionally, planks will show a natural upward bow in the middle. Sometimes you can minimize this condition by back scoring or kerfing the back of the plank perpendicular to its length on a table saw. Be sure you do not saw or kerf too deeply into the back so that it mars the surface. You may also apply weights overnight on bowed sections in order to allow the adhesive to dry and develop a stronger bond.

7. It may be necessary from time to time to align the boards. Use a cut-off piece of scrap as shown in Fig. 7-17 or use a rubber mallet to tap boards on the face until moved into position. Align the boards at the wall line with a pry bar.

8. When the short area is completed, finish off the main work area in the same manner, and cut the last row to fit at the wall, leaving a 1/2" expansion space. To overcome any appreciable difference in elevation between the new plank floor and the floors in adjoining rooms, most flooring manufacturers have a special 3/8" by 1-1/2" nosing strip available.

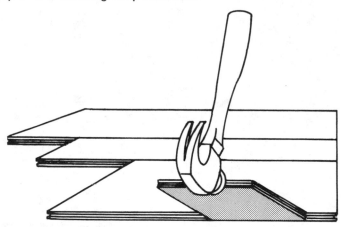

Fig. 7-17: Aligning the planks.

Strip or Plank Flooring Over Concrete

While there are several ways to lay strip or plank flooring over concrete, including the one described on page 125, the cost-cutting method described here does not use a wood subfloor, but does meet Federal Housing Administration (FHA) requirements. The technique involves use of a double layer of 1" by 2" wood sleepers nailed together, with a moisture barrier of 4 mil polyethylene film between them. The bottom layer of sleepers is secured to the slab by mastic and concrete nails. The strip hardwood flooring then is nailed at right angles to the sleepers, with one nail at each bearing point (Fig. 7-18). The five steps of installation are as follows:

1. **Applying the Adhesive.** Make sure that the floor is clean, dry, and free of loose concrete scale or paint. Snap chalk lines on 16" centers parallel with the longest wall. Apply a good grade of construction adhesive in a serpentine bead approximately 3/8" thick on these lines. Then, position 1" by 2" furring strips

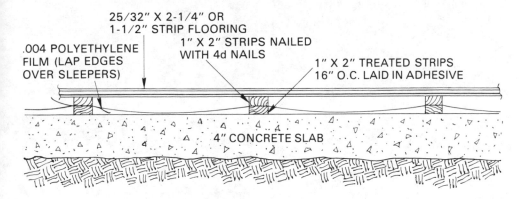

25/32" X 2-1/4" OR
1-1/2" STRIP FLOORING
1" X 2" STRIPS NAILED
WITH 4d NAILS

.004 POLYETHYLENE
FILM (LAP EDGES
OVER SLEEPERS)

1" X 2" TREATED STRIPS
16" O.C. LAID IN ADHESIVE

4" CONCRETE SLAB

Fig. 7-18: Laying strip floors over a concrete slab.

over the center of the line and on approximate 16" centers. If heating coils are in the slab, make sure that the adhesive will tolerate the temperatures that are normally encountered.

2. **Installing the Bottom Sleepers.** The bottom sleepers should be 1 by 2's treated with wood preservative. Imbed the strips in construction adhesive, making certain that they are positioned properly on 16" centers. These 1 by 2's should be random lengths laid end to end with slight spaces between the ends not butted tightly together. Concrete nails can also be used in conjunction with the adhesive, but they are not necessary. Caution should be used when driving concrete nails, and safety glasses should be worn at all times. Best results are obtained if the adhesive bonding the sleepers to the floor is allowed to dry for approximately 24 hours before proceeding with the installation.

3. **Applying Polyethylene Film.** After all the bottom sleepers have been installed, 4 mil polyethylene film should be laid over the first course of strips, joining polyethylene sheets by lapping edges over sleepers.

4. **Laying the Top Nailing Sleepers.** The second course of 1 by 2's (which do not have to be preservative treated) should be nailed with 4d nails, 16" to 24" apart. Nails shall go through the top sleeper, polyethylene, and into the bottom sleeper.

5. **Installing Strip or Plank Flooring.** Install the flooring at right angles to sleepers by blind nailing to each sleeper, driving at an angle of approximately 50°. Nails should be threaded or screw type, cut nail or barbed fastener. No two adjoining flooring strips should break joints in the same sleeper space. Each strip should bear on at least one sleeper. Provide a minimum of 1/2" clearance between flooring and wall to allow for expansion. Do not attempt to install laminated wood plank in this manner.

RESILIENT FLOORING

Practicality, convenience, ease of application, and long wear are a few of the reasons why resilient flooring continues to be one of the most popular choices of homeowners everywhere. With the tremendous variety of materials, designs, and colors available, it is possible to create just about any floor scheme that strikes your fancy (Fig. 7-19). For example, you can install a floor of all one color, or you can combine different colorings into a custom floor design that matches room requirements and individual taste. If your decorating taste leans to the natural look, you will find countless resilient materials that closely resemble the

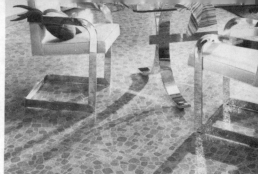

Fig. 7-19: Examples of available designs of resilient flooring (Courtesy of Armstrong Cork Company).

appearance of slate, brick, wood, terrazzo, marble, and stone. Many of these floors feature an embossed surface texture that adds a striking note to the design.

Types of Resilient Floors

Resilient floors are manufactured in two basic types: (A) sheet materials and (B) tiles. The latter are cemented in place to serve as a permanent floor. Sheet materials are also cemented in place, but in some cases can be installed loosely like rugs. Tiles generally come in 9" or 12" squares; sheet materials are available in continuous rolls up to 12' wide.

On the next page is a rundown of the various resilient floor coverings and their performance characteristics.

Tiles. Made of vinyl, asbestos fibers, and other components, resilient tiles are exceptionally durable and easy to keep clean. They lend themselves to a variety of customizing effects, since tiles of different colors and styles may be easily combined. In addition, most tile floors are ideal for do-it-yourself installation.

What kind of tile should you select? Asphalt tile, the first resilient tile, is the least expensive and can be installed at any grade level. It offers good durability, but compared to other types of resilient floors, it ranks low in resistance to grease and soil. Most asphalt tiles are 1/8" in thickness.

Vinyl-asbestos tile is the most popular of all resilient tiles. It is inexpensive and can be installed nearly anywhere, above, on, or below grade. Vinyl-asbestos tiles have exceptional durability and are easy to clean. They do not require waxing; they can be given a low sheen by buffing after the floor is mopped. Wear resistance is generally rated very good. These tiles are available in 1/16", 3/32", and 1/8" thicknesses.

Solid (or homogeneous) vinyl tiles that have a backing are the ultimate in the tile type of flooring. They rank as excellent in durability and have a surface that is smooth and nonporous. This makes upkeep easy and economical. You can use

Material	Backing	How installed	Where to install	Ease of installation	Resilience and maintenance	Durability	Quietness
TILE MATERIALS							
Asphalt	None	Adhesive	Any interior	Fair	Difficult	Fair	Very poor
Vinyl asbestos	None	Adhesive	Any interior	Easy	Very easy	Excellent	Poor
Vinyl	None	Adhesive	Any interior	Easy	Easy	Good—excellent	Fair
Cork	None	Adhesive	On or above grade	Fair	Fair (with vinyl, good)	Good	Excellent
SHEET MATERIALS							
Inlaid vinyl	Felt	Adhesive	Above grade	Fair	Easy	Good	Fair
	Foam and felt	Adhesive	Above grade	Difficult	Easy	Good	Good
	Asbestos	Adhesive	Any interior	Very difficult	Easy	Excellent	Fair
	Foam	Adhesive	Any interior	Very difficult	Easy	Excellent	Good
Printed vinyl	Felt	Loose-lay	Above grade	Easy	Fair	Poor	Poor
	Felt	Adhesive	Above grade	Fair	Easy	Fair	Poor
	Foam and felt	Loose-lay	Above grade	Easy	Easy	Fair	Good
	Foam and asbestos	Adhesive or loose-lay	Any interior	Fair—easy	Easy	Good	Good
	Foam	Loose-lay	Any interior	Easy	Easy	Good	Good

them on any grade level. Solid vinyl is available in many colors and patterns, ranging in thickness from .080 gauge (thin) to 1/16" and 1/8".

Cork, possibly the quietest of all floorings, is available as pure natural cork or combined with vinyl. The latter combination retains the beauty and warmth of cork while providing an added degree of cleaning ease. Vinyl cork has a higher degree of durability than natural cork.

Sheet Flooring. The principal advantage of sheet flooring is seamlessness. Since it is installed in wide rolls, there are few seams in the finished floor. The result is a beautiful wall-to-wall sweep of color and design, a perfect setting for room furnishings. Some sheet floors can also be customized by combining two or more colors or styles. Sheet vinyls are resilient, therefore comfortable underfoot, and resistant to grease and alkalies as well. With special backing they can be used below, on, or above grade. Resistance to wear is generally very good.

Preparation for Resilient Floors

The preparation of the subfloor for installation of resilient tiles or sheet flooring is essentially the same. If the present flooring is perfectly smooth, level, and in good physical condition, the new resilient floor may be applied directly to it. Scrub the floor thoroughly to make sure that it is completely free of wax, paint, varnish, oil, or grease. Fill all the cracks and depressions with wood plastic or filler. Sand down any high spots and renail any loose boards of wood floors. Scaling concrete should be chipped out and patched with concrete bonder or a concrete latex underlayment made specially for that purpose.

When the APA glued floor system (described earlier in this chapter) is used in new construction, the resilient floor coverings can be applied directly to the plywood subfloor.

It is wise, if there is any doubt in your mind regarding the condition of the old flooring, to install an underlayment. Also, in new work, resilient floors should not be installed directly over a board or plank subfloor. An underlayment grade of woodbased panels, such as plywood, particleboard, and hardboard, is widely used for suspended floor applications. But, when buying underlayment, be sure to specify underlayment for resilient flooring. For instance, do not use ordinary hardboard. Tempered hardboard is too hard. It is difficult to drive nails through it and get the heads flush with the board. Protruding nailheads will produce a bumpy, uneven floor. Untempered hardboard is too soft and doesn't provide a firm base for tile. Sharp objects, especially a woman's high heel shoe, are liable to punch a hole in the material. Underlayment grade hardboard is sold by most floor tile dealers. Ordinary hardboard is sold at lumber yards.

The underlayment grade of particleboard is a standard product and is available from many producers. Manufacturer's instructions should be followed in the care and use of the product. Plywood underlayment is also a standard product and is available in interior types, exterior types, and interior types with an exterior glue line. The underlayment grade provides for a sanded panel with a C-ply or better face ply and a C-ply or better immediately under the face. This construction resists damage to the floor surface from concentrated loads such as chair legs and tables.

Do not install resilient flooring directly over old double wood floors that are badly worn, have loose or broken boards, or have top boards 4" or more in width. Instead, you should first nail down loose boards, replace broken ones, and repair all small holes or cracks by using wood patching material. Then, if the top boards are less than 4" wide, install 1/4" underlayment grade plywood or 1/4" hardboard underlayment board, smooth side up. Where top boards are 4" or more in

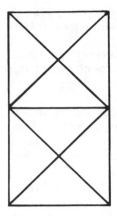 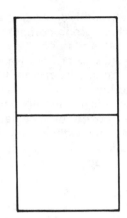 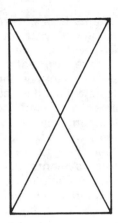

Fig. 7-20: Bead pattern for installing underlayment.

width, or over single wood floors, 3/8" or 1/2" underlayment grade plywood should be used.

The underlayment is best installed by using an adhesive-nail combination. Apply a panel or construction adhesive to the back of the underlayment, using one of the bead patterns shown in Fig. 7-20. After the back surface has been applied to the old flooring surface, nail the underlayment every 3" to 5" along the edges and every 6" to 8" throughout with underlayment nails.

When installing underlayment, there are two points to keep in mind. First, see to it that the panels are **not** butted tightly together. Leave a space between them equal to the thickness of a pencil point or matchbook cover (Fig. 7-21). This allows for expansion. Second, stagger the panels to avoid having four corners meet at one spot. Tile placed on top of the corners may shift.

Resilient flooring can also be installed over existing resilient floor coverings (not embossed) of smooth surface vinyl sheet flooring (except material with a foam back or interliner) linoleum, cork, rubber, solid vinyl, asphalt or vinyl asbestos tile with the following exceptions:

1. Asphalt tile is not recommended for installation over existing linoleum or cork tile at any grade level.

2. Vinyl asbestos tile is not recommended for installation over existing cork tile. In addition, the existing floor covering must be presently installed over a suspended floor—meaning an underfloor which is *not* in contact with the ground.

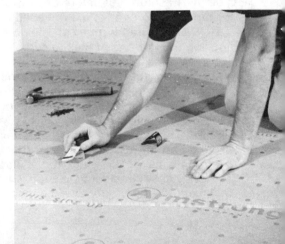

Fig. 7-21: Allow a little less than 1/32" (the thickness of a paper matchbook) between each 4' by 4' panel of underlayment or plywood to permit expansion.

Existing floor covering (securely bonded) of vinyl asbestos, solid vinyl, asphalt tile, and vinyl sheet flooring which are in contact with the ground may be covered with asphalt or vinyl asbestos tile only.

Make sure that all traces of old wax or floor finish are entirely removed from the surface to be covered. This should be accomplished by thorough scrubbing, using a vinyl floor cleaner together with an abrasive scouring powder. Also, be sure that the old floor covering is firmly adhered to the underfloor. Sections of linoleum which may have come loose should be nailed down, flush to the underfloor, with underlayment nails. Any badly worn spots in the existing floor covering should be leveled with a good wood filler. In some cases, the concrete latex underlayments will bond equally as well to wood and provide a good leveling compound.

At the time you are preparing the underfloor, you might want to consider prying up the existing molding at the base of the walls so you can run your tile underneath it for a neater finished appearance. Moreover, if you are planning to install vinyl wall cove base (page 149), you will want to completely remove existing baseboard moldings at this time.

Resilient Flooring Over Concrete. Concrete for resilient floors should be prepared with a good vapor barrier installed somewhere between the soil and the finish floor, preferably just under the slab. Concrete should be leveled carefully when a resilient floor is to be used directly on the slab, to minimize dips and waves. Smooth off any rough spots and fill any crack or level the surface with concrete bonder or concrete latex underlayment used as described on page 171. Oil base paint on concrete in direct contact with the ground must be removed with a floor sanding machine with No. 4 or No. 5 sandpaper (open coat). Such sanders can generally be rented at your local hardware store. It is unnecessary to remove rubber base paint if it is in good condition. Most modern resilient floor products can be cemented directly to concrete without the use of primers, sealers, hardeners, or felt lining.

Before laying any resilient material on concrete, be sure that all traces of wax, grease, and dirt are thoroughly removed to insure proper bonding of the adhesive. Never install tile over concrete that is wet or damp. Use the simple test to determine the presence of moisture outlined earlier with a 4' by 4' sheet of polyethylene (see page 135).

Installing Floor Tile

Laying the right kind of tile so that it looks neat and will stand up over the years requires careful planning. The various previously mentioned types are all laid in a similar fashion, but before any installation work can be started, you must determine the number of tiles needed and how they should be arranged.

How to Figure Number of Tiles Needed. The following table will aid you in figuring the number of tiles to complete an installation job. For instance, if you are working with a floor area which is 280 square feet (a 14' by 20' family room), and you want to use 9" by 9" tiles, the table indicates 356 tiles for 200 square feet and 143 tiles for 80 square feet, a total number of 499 tiles.

When ordering tiles, it is most important to consider the waste factors. In our example, the allowance for waste is 7% of the total number of tiles, or an extra 35 tiles. This would make a grand total of 534 tiles. Since tiles are usually boxed 80 to a carton, this would mean that we need over 6-3/4 cartons. Even if the dealer is willing to split a carton, it would be wise to take the seven full cartons. This will assure an adequate supply of tiles from the same lot and also allow for replacement if they are ever needed.

Square feet	Number of tiles needed (inches)			Square feet	Number of tiles needed (inches)		
	9 x 9	12 x 12	9 x 18		9 x 9	12 x 12	9 x 18
1	2	1	1	60	107	60	54
2	4	2	2	70	125	70	63
3	6	3	3	80	143	80	72
4	8	4	4	90	160	90	80
5	9	5	5	100	178	100	90
6	11	6	6	200	356	200	178
7	13	7	7	300	534	300	267
8	15	8	8	400	712	400	356
9	16	9	8	500	890	500	445
10	18	10	9	600	1,068	600	534
20	36	20	18	700	1,246	700	623
30	54	30	27	800	1,424	800	712
40	72	40	36	900	1,602	900	801
50	89	50	45	1,000	1,780	1,000	890

Allowance for waste

1—50 square feet	14 percent
50—100 square feet	10 percent
100—200 square feet	8 percent
200—300 square feet	7 percent
300—1,000 square feet	5 percent
Over 1,000 square feet	3 percent

Laying Out The Pattern. As was stated earlier in this chapter, one of the major advantages of most tile materials is that they lend themselves to a variety of customizing effects. For instance, feature strips, from 1/2" to 3" wide, are available and can be employed to border a room or outline individual tiles, or they can be laid diagonally in a herringbone pattern; both solid and variegated colors are used. The use of insets is another way to give your floor an individual look. These are available in several different picture designs, and one is sure to fit the theme of your room. You can also choose a standard three-letter monogram for your inset design. But, before laying out any pattern, you must mark the starting point, the center of the room. This is accomplished in exactly the same manner as for wood tiles or blocks described on page 128. Remember that before spreading the adhesive, lay a row of loose tiles along the chalk lines, starting at the center point and working out to one side wall and then to the other end wall. This determines the space left for the outside row of tiles (the border tiles). Measure the distance between the wall and the last tile. If the distance is less than 4" or more than 10" (for 12" square tiles), move the center line parallel and 6" closer to that wall. This will prevent the peculiar look of tiles that are too small along either wall. If the central point is moved and either line is resnapped, check both lines again to be sure that they are at perfect right angles to one another. **Note:** If your floor design makes use of a feature strip, be sure to include it at the appropriate intervals in the dry run, since the amount of feature strip used will have a bearing on the width of your border tiles.

Laying the Tile. Before spreading the adhesive, be sure to thoroughly read the detailed directions given on the label of the adhesive container. *This is extremely important,* since different adhesives require different methods of application. For instance, some adhesives can be applied with an ordinary paintbrush or short-nap type paint roller, while others *must be* spread with a properly notched trowel. A few multipurpose adhesives can be spread by brush, roller, or trowel. The amount of adhesive spread at one time and the waiting period required before installing the tile are other variable factors that will depend upon the adhesive specified for the floor you are installing. You should begin spreading your adhesive at the intersection of the guide lines, working one quarter of the room at a time. As you proceed along in accordance with the directions on the label, spread the adhesive right up to the working line, but be careful not to obliterate it by spreading adhesive over it (Fig. 7-22). Also, when spreading the adhesive, remember that too much cement "bleeds" through between tiles and also forms a soft coating under the tile which permits heavy objects to dent the tile. Too little adhesive results in loose tiles and possible cracking.

As soon as the recommended adhesive is ready to receive tile, you should place the first tile to be installed exactly in the right angle formed by the chalk lines so that the two sides of the tile are precisely on the guide lines. Place each tile firmly against the tile already in position and lower the tile into the adhesive. Keep tiles on the guide line and have corners meet exactly. Do not slide tiles into position, as this causes adhesive to ooze up between tiles. When installing marbleized and certain other patterns, it is preferable to alternate the direction of veining in the adjacent tiles.

Fig. 7-22: Laying resilient tiles.

Once the first section is completed, spread the cement in the second quarter of the room and put down the tile in that quarter. Then, do the third and fourth quarters. Whenever you start a new quarter, be sure the tiles along the chalk lines are tightly and accurately positioned against these straight and squared lines.

When installing asphalt or vinyl asbestos tile, you should work out from the center lines by kneeling directly on the tile, pyramiding out with each succeeding row of tile. Never roll asphalt or vinyl asbestos tile.

When installing solid vinyl, rubber, or cork tiles, you should be sure to kneel only on the underfloor and not on the newly cemented tiles or on the adhesive. When you run out of clear floor space, do not kneel directly on freshly cemented tiles. Use a piece of 2' by 2' hardboard or plywood as a kneeling board. Also, when you are installing any solid vinyl, rubber, or cork products, be sure to roll the installed tile immediately and thoroughly. With each additional 18 tiles laid, again roll the previously installed tiles, and repeat this rolling procedure thoroughly as the installation progresses. Floor rollers may be borrowed or rented from your dealer, or if not available, use a common rolling pin. Be sure to roll completely, exerting pressure to make certain the back of the tile makes a good bond with the adhesive.

If your floor design calls for feature strip, just remember that it should be installed exactly as you would a piece of tile, i.e., place it firmly against the tile already laid and lower it into the adhesive. Do not slide it into position. Always lay sufficient tile so that you have tile adjoining the full length of feature strip. This will eliminate the possibility of having a portion of the feature strip extending out into the adhesive. Also, when you are working with feature strip, be sure to have the floor design worked out in advance so that you will know just where each row of feature strip should fall.

After installing the field tile in each quarter of the room, you can proceed to cut-in your border tiles. Place a loose tile over the last row of installed tile, taking care to see that the graining is running in the right direction for the border. Then, on top of this loose tile, place another tile flush against the wall, placing a strip of wood in the adhesive to keep the tile clean. Using the topmost tile as a guide, mark the underneath tile with a pencil or sharp awl. When trimmed, it will fit perfectly in the border area (Fig. 7-23A). Figures 7-23 B and C show the steps in marking a corner tile. The following pointers will prove helpful to you in cutting and fitting the various types of tile:

1. **Asphalt Tile.** For straight cuts, asphalt tile should be scored deeply with a sharp knife or awl. The tile will then snap off along this line, starting at one end of the tile—never in the center. Smooth off the uneven end of the tile with sandpaper. For irregular cuts, tile should first be heated until pliable and then cut with a sturdy pair of scissors.

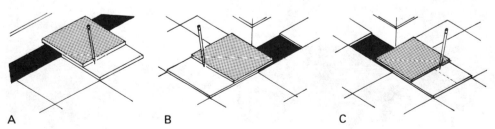

A   B   C

Fig. 7-23. Marking for a border tile and corner tile.

2. **Vinyl Asbestos Tile.** Eighth-inch gauge tile should first be heated with a heat lamp until pliable. A pair of sturdy household scissors or a utility knife is all that is needed to cut this product.

3. **Vinyl and Rubber Tiles.** These products can be cut with a pair of sturdy scissors or a utility knife after they have been warmed in an oven, on an electric food warmer, or held in front of a portable electric heater. It is not necessary to get the tile hot—only well warmed. Remember, you do not want to burn your fingers or the tile.

4. **Cork Tiles.** These tiles can be cut with a utility knife.

There are some locations where the edge of the tiles will not be covered (as with base shoe molding around most of the room), and they must be scribed exactly. In such areas, place the tile in its true position relative to the position it will finally occupy. It must be parallel to the field tile (or already laid tile), but not necessarily to the wall, which might be irregular. Then, scribe the lines perpendicular to the field and out from the wall with an extra tile or a steel square with one edge true to the field. Use dividers for marking the distance from the piece being fitted. Such a scribing procedure is usually necessary when fitting tiles exactly around door jambs. Perhaps, the neatest way of all is to make a simple saw cut at the base of the door frame. Remember, the door jamb is not a structural support, and you can easily take a carpenter's saw and make a quick saw cut to remove about 1/8" of the frame at the base. You can then slide your border tile up tightly against the wall and completely eliminate the need for cutting the tile around the door frame. Another simple trick for achieving a neater installation is to completely remove any unsightly door saddles and run your tile straight through the doorway instead of cutting and fitting it around the saddle.

To fit tiles around pipes (Fig. 7-24) or other obstructions, make a paper pattern to fit the space exactly. Then, trace the outline onto the tile and cut it accordingly.

Fig. 7-24: One way of fitting a tile around a post, pipe, or similar obstruction. Place the tile against the side of the post, mark the width, and repeat on the other side of the post. Draw the cut mark; once cut, the tile can be put in place.

For an intricate pattern, it may be necessary to heat the back of the tile over a floor or heat lamp to make it more pliable. Toilets and other large obstructions are best removed and tiled underneath. Be sure to disconnect the water when doing this.

By setting tiles carefully, you should be able to avoid excess adhesive coming up between the tiles. Some excess adhesive is bound to get onto the tile surface, however, and it should be removed immediately. Follow the directions on the cement container to wipe up the cement. Be careful not to allow solvent to run down between the tiles, as it will loosen them.

Do not wash or wax the floor for five to ten days after installation, until the tiles have become thoroughly bonded to the subfloor. Sweeping them with a soft broom or cleaning them with a damp cloth or mop is the only maintenance necessary during this period.

Waxing is not necessary with most resilient floor tile unless desired. Mopping and buffing will usually keep the floor looking new and attractive. But, if waxing is desired, be sure the floor is thoroughly cleaned. Apply a good floor wax in a very light coat. Apply the wax in straight motions (not sweeping or circular). For a higher shine, buff the floor by hand or use an electric polisher. Do not wax over dirty floors. Excessive wax produces areas where wax film is too thick and usually causes a floor to look discolored.

All resilient floor coverings will indent, more or less, if weight is applied to a small, concentrated area. Therefore, remove all small metal domes or buttons from furniture legs. Use broad, flat glides or cups. Replace hard, narrow rollers with soft wide rubber rollers.

Replacing Damaged Tiles

Breaks and dents occasionally occur in resilient tiles. To remove a damaged tile, heat it with a light flame from a propane torch, or use an electric iron covered with a towel. When employing a torch, go completely over the tile with the flame, using an asbestos board at the joints to avoid damage to the adjoining tiles. Then, with an old chisel, break out the damaged tile, keeping the bevel of the chisel down. You may have to reheat areas as you go. It is always a good idea to keep a fire extinguisher handy when using a torch in this manner. Use extreme caution so that you do not have the flame in contact with the floor for too long a period or that you do not scorch the subfloor.

Clean the subfloor and make any patches, if necessary. Sand the subfloor smooth and level so there are no lumps or holes. Then, butter the back of the replacement tile with floor tile adhesive. Spread it fairly thin—especially at the edges. Butt the new tile against the adjoining one and drop it into place. Then, press it down firmly to ensure a smooth and level bond.

Sheet Vinyl

Sheet vinyl flooring—either inlaid or printed—used to be a source of trouble to do-it-yourselfers. Today, new materials, adhesives, and methods make installing sheet goods almost as easy as putting down tiles.

Modern sheet vinyl flooring is light, easy to both handle and cut, and so flexible that it can be bent at a sharp angle without cracking. Available in 6', 9', 12', or 15' widths, it can be installed without seams in most rooms.

It is a good idea to make a sketch of the room, showing alcoves, bays, counters, or other features that will affect the shape and size of the flooring (Fig. 7-25). Never apply any adhesive until all cutting and piecing is complete and you are sure that the flooring fits the layout. Lay out the flooring material in a convenient larger room, such as your garage or basement. Transfer the outline of the room to the flooring material adding 3" to all measurements (Fig. 7-26). Cut out with a

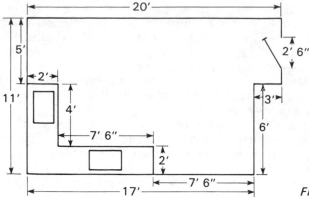

Fig. 7-25: Typical room sketch.

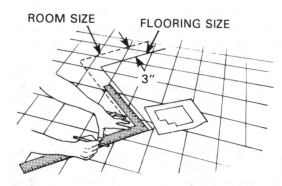

Fig. 7-26: Transferring measurements to the sheet material.

sharp utility knife or heavy scissors or shears. If you use a utility knife, be sure that you provide protection against the cutting of any finished flooring below.

Transfer the flooring material, rolled pattern-side up, into the room in which it will be installed. When the material is positioned, it will be possible in many rooms to lay the factory edge of the material against a long wall, provided it is straight and square with the other walls. This will save cutting of the flooring along this wall, besides acting as a guide to aligning the pattern of the material in the room. One caution—be sure that the floor molding to go back on top of the material will cover any edge markings in the floor covering. For the best results and easiest installation, the room should be warm. The subfloor, flooring material, and room temperature should be at least 70°F.

After the floor material is positioned in the room, make a relief cut at the outside corners from the top of the lapped up material to the point where the floor and wall meet (Fig. 7-27A). Then, on the inside corners, gradually cut down the flooring material a little at a time until it fits snugly into the corner (Fig. 7-27B). Next, using the free-hand method, take a sharp utility knife and gradually trim down the flooring material lapped up the wall until it lays flat (Fig. 7-27C). The maximum allowable gap between the wall or baseboard and the edge of the flooring material should be about 1/8" less than the thickness of the molding that is to be reapplied after the flooring is installed.

Another method of trimming down the material lapped up the walls is to press it into a right angle where the floor and wall meet. Use fingers or a piece of 2" by 4" (Fig. 7-28A). Then, lay the metal straightedge as close as possible to the wall and cut along the edge with a sharp utility knife (Fig. 7-28B).

Fig. 7-27: Fitting to the wall.

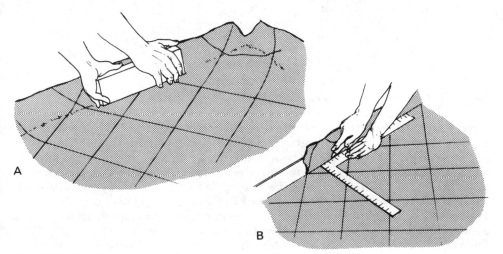

Fig. 7-28: Another method of trimming down the material.

Since doorways will receive no molding, they should be fitted flush. The best method to get a good fit is to make a series of vertical slits in the lapped up flooring material to the subfloor. These slits should be made at all outside and inside angles of the doorway (Fig. 7-29A). Next, take a screwdriver or other piece of flat metal and press the floor covering sharply into the right angle formed by the subfloor and doorway trim (Fig. 7-29B). Hold the utility knife (blade must be sharp) at an approximate 45° angle and trim off the excess flooring material (Fig. 7-29C).

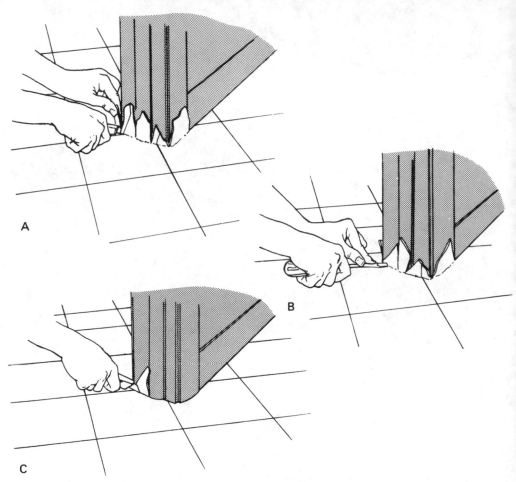

A

B

C

Fig. 7-29: Fitting around a doorway.

It is not feasible to try to fit long distances by free-hand trimming, as described earlier, unless a molding is used to cover the cut edge. Occasionally, it may be desirable to fit the floor covering snugly against straight walls or objects, such as kitchen cabinets. The best method for areas without molding is to position the flooring material a few inches from the object to be fitted. Take an accurate measurement from the wall or object and mark the surface of the floor covering every 15" to 18". Connect the marks with a straightedge to make an exact line of the object you are fitting against. Cut along the straightedge with a sharp utility knife. Slide the flooring material flush against the object which is being fitted, and then cut in the balance of the room by the free-hand trimming method described earlier.

In instances where the wall to be fitted is irregular, such as one with cabinets and a refrigerator alcove, this same method of fitting can be used. Position the flooring material flat on the floor and as close as possible to the farthest point to be fitted. Measure that distance. Then, mark the flooring material so it will fit to the farthest point. Using the same measurement, accurately mark the flooring along all parallel sections of the cabinets or wall every 15" to 18". Lay the straightedge along each end of the cabinets and extend the lines to form right angles, completing the alcove outline. Using the straightedge as your guide, cut out the outline and slide the flooring into place. The balance of the room should be trimmed-in, using the free-hand method.

Completely fit the floor covering to the room. Then, roll the material, pattern-side in, or lay it back on itself, so that about one half of the subfloor is exposed. Using a trowel, apply vinyl floor adhesive on the subfloor. Do not apply adhesive too far ahead. Some adhesives dry very rapidly and will not develop an adequate bond after a certain period of time. It is always a good idea to occasionally pause and lift up a slight edge of the flooring to make sure there is still adequate adhesive transfer from the floor to the back of the flooring material being installed. Then, roll the flooring material back onto the adhesive. Take an ordinary push broom and completely flatten the flooring material, being certain that all air pockets are expelled. Repeat the same procedure on the second half of the room. Replace the moldings and cover the edges of the floor covering at the doorways with metal trim.

Although sheet vinyl flooring is manufactured in widths up to 15', it may occasionally be necessary to seam two or more sheets in very large rooms. Your retailer will advise you, at the time of purchase, of the additional material needed for matching the design you select. Many designs require reversing of sheets at the seam. He will also guide you on this.

Fit the first sheet to the room. Apply the adhesive to within 8" to 9" of the seam area. Flatten the sheet against the subfloor with a push broom. Overlap the second sheet at the seam so the design unit matches across the sheets. Carefully determine that the design on the top and bottom sheets aligns exactly before proceeding further. If the sheets are not exactly aligned, the design will not match when the seam is cut.

After matching the design, fit the second sheet to the room and apply adhesive to within 8" to 9" of the seam area. Push broom flat against the subfloor as on the first sheet. Following a metal straightedge, cut through both layers of flooring material where they overlap, with a very sharp utility knife held straight up. Turn back both edges of the flooring and apply the adhesive to the seam area. Then, push broom the flooring completely flat against the subfloor. Clean off excess adhesive with a clean damp sponge or cloth. Be certain that both sides of the surface are level with each other at the seam.

Wall Cove Base Installation

The dirt catching gap where the wall meets the floor is still a trouble spot in many homes. Dirt collects there and spreads to other parts of the floor, and it takes real hands-and-knees scrubbing to keep this "drudgery zone" clean.

The answer to needless work is the installation of a wall cove base. Available in the popular resilient floor materials—rubber, vinyl, and asphalt—and a range of attractive basic colors, it provides a smooth curved surface and a tight joint where the wall meets the floor. Cove base never needs painting, cleans easily, and it stays bright and new looking. It is usually produced in 2-1/2", 4", and 6" heights and 48" and 96" lengths.

You can easily apply wall cove base over any smooth, dry, clean wall that is not in contact with the earth. The corrugated back grips tightly. The tapered top and coved base make a perfect dust-tight seal. Be sure that the wall surface is free from oil, grease, loose paint, or other foreign matter. The wall surface must be continuous all the way down to the floor, without large gaps at the bottom. Do not install cove base on outside walls in contact with the earth, such as basement walls. These walls should be furred out. Do not install it over vinyl covered walls or over paint less than two weeks old.

To install the wall cove base, proceed as follows:

1. Apply the cove base adhesive to the back of the cove base with a spreader having notches approximately 1/16" deep and 1/16" wide or a putty knife. When spreading the adhesive (Fig. 7-30A), leave 1/4" bare space along the top edge of the base so that the adhesive does not ooze above it. Adhesive smears on the surface, as well as tools, should be removed immediately with a cloth dampened in alcohol.

2. Press cove base firmly and uniformly against the wall surface within 10 minutes after applying adhesive. Make sure the toe is tight down to the floor.

3. Roll the entire surface of the base with a steel hand roller. If there is no hand roller available, use any smooth, clean cylindrical object (Fig. 7-30B). Exert pressure so that the cove base will adhere to the wall at all points. After rolling, press

A

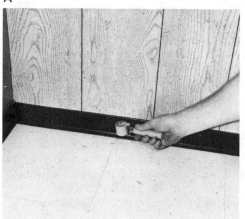

B

C

Fig. 7-30: Steps in installing a wall cove base.

the toe of the base firmly against the wall with a straight piece of wood (Fig. 7-30C). Irregular wall surfaces or curved wall surfaces may require bracing until the adhesive has set.

Corners. Preformed inside and outside cove base corners are available in all colors and heights. However, it is easy to make your own inside and outside corners by following these directions: The easiest method is to make and apply the corner first. Then fit the straight sections to it. For either an inside or outside corner, begin by carefully marking the base where the corner will be made.

To make outside corners, proceed as follows:

1. Heat the back of the base over a hot plate until it is soft and pliable. Be careful not to scorch the face.

2. While still hot, fold the cove base *back-to-back* at the corner mark and roll exactly over the mark with a hand roller or rolling pin (Fig. 7-31A).

3. Next, hold the hot cove base tightly folded in a pail of water to cool. Cooling while tightly folded gives the cove base a built-in "spring" that grips the corner.

4. Unfold the base, wipe it dry, and apply the adhesive. Then, press it into place and roll (Fig. 7-31B).

A B

Fig. 7-31: Installing an outside corner.

To make inside corners, proceed as follows:

1. Where the corners will occur, cut out a triangular wedge in the cove base toe with a sharp knife (Fig. 7-32A).

2. Apply the adhesive and press the cove base into the corner with a piece of tile or a straight piece of wood (Fig. 7-32B). The pressing action closes up the wedge in the corner.

A B

Fig 7-32: Installing an inside corner.

An alternate method for installing cove base is to use a contact cement. First, position the cove base against the wall and lightly draw a pencil line along the top edge of the cove base. Remove the cove base and apply a brush coating of contact cement on the back of the cove base and directly to the wall in the area below the pencil mark. Allow the contact cement to dry in accordance with the manufacturers directions. Then, carefully position the cove base against the wall and press firmly into position. Remember that you cannot make any adjustments when contact cements are used, as the bond is immediate and positioning must be exact.

RIGID OR HARD FLOOR MATERIALS

In special locations within the home, rigid or hard floor materials, such as ceramic tile, slate, flagstone, and brick, may be used. In fact, ceramic tile is one of the oldest floor-covering materials; it dates back almost 7,000 years.

Ceramic Tile Floors

There are three types of ceramic tiles (Fig. 7-33) in common use for floors today: quarry tiles, ceramic mosaics, and glazed tiles. The latter are usually a little thinner than glazed wall tiles, but are made in various sizes and shapes and a variety of designs and colors. Some are so perfectly glazed that they form a monochromatic surface. Others have a softer, natural shade variation within each unit and from tile to tile. In addition, ceramic floors can be bright-glazed, matte-glazed, or unglazed. There are also extra-duty glazed floor tiles suitable for heavy-traffic areas.

Ceramic mosaics are available in 1" by 1" and 2" by 2" squares and come with or without a glaze. In addition to the standard units, they may be had in a large

Fig. 7-33: Three types of ceramic tiles: (left to right) quarry tiles, ceramic mosaics, and glazed tiles (Courtesy of American Olean Tile Company).

assortment of colorful shapes. Mosaics are usually sold mounted on a webbing in 1' by 1' and 1' by 2' sheets for easy installation.

Quarry tiles, which are also made from natural ceramic materials, are available in a variety of colors; the most common types are in shades of red, chocolate, and buff. They come in shapes ranging from square tiles to Spanish forms.

Laying Ceramic Floor Tiles. At one time it was thought that the only base on which floor tile could be laid was a heavy layer of concrete made of three parts sand and one part cement. The base was made about 3" thick and reinforced with wire mesh. Although some such construction is still desirable in public buildings where the floor must bear heavy traffic, the tendency in houses has been toward a less bulky and lighter construction. With the heavier construction, the floor joists are partly cut away, boards are fastened between the joists a couple of inches below the tops, waterproof building paper is laid on the boards, and concrete is poured on top of that. The tile is then laid on top of the concrete.

Today, ceramic floor tiles can be applied to almost any surface that is in good condition, firm, level, perfectly smooth, and free from moisture and foreign matter. This includes double wood flooring, 1/2" exterior type plywood, ceramic tile, steel-troweled cement, and asphalt or vinyl tile, or sheet resilient flooring. Do not lay tile over a springy floor. Nail down loose flooring. If this does not correct the situation, cover the floor with an exterior grade plywood underlayment at least 3/8" thick before tiling. Allow 1/8" expansion joints between the plywood edge and wall. For a floor that needs leveling, apply a coat of concrete bonder or suitable concrete latex underlayment as described on page 171.

Ceramic tiles are laid out in the same manner as the wood parquet type described on page 128. That is, snap a chalk line from wall to wall. Then, loose lay the tiles to make sure you do not wind up with less than half a tile at the edge. Now, snap another chalk line between the other two walls. Using a square, check that the lines cross each other at right angles. Adjust the starting point so that you do not wind up with less than half a tile at the outer edges.

Apply the floor tile adhesive according to the manufacturer's directions on the container. Be sure that you select an adhesive specially formulated for the particular flooring material you are installing. Not all ceramic floor tiles require the same type of adhesive and, in many cases, even require a different sized spreader because of the thickness or size of the tile. Use a notched trowel to spread it over a 3' square area of the floor (Fig. 7-34A). Do not cover any lines that you have drawn on the floor for alignment or reference. If you are installing extra thick quarry tile with a deep back pattern, it is usually a good idea to also spread adhesive on the back of each tile.

Set the tiles by starting at the reference lines near the middle of the room, and work toward the walls. Carefully yet firmly press each tile into place (Fig. 7-34B), twisting it slightly with your fingertips to set it well into the adhesive. Do not slide the tiles against each other, or you will wind up with excessive adhesive build-up in the corners. In most installations, use pieces of 1/16" thick cardboard or wood as spacers to keep the joints between the tiles to a uniform size; most floor tiles do not have spacer lugs (Fig. 7-34C).

From time to time, check with a straightedge or level and square that the tile courses are true. To make sure that the tiles are flat and firmly embedded in the adhesive, take a piece of 2" by 4" lumber about the length of three tiles, and pad it with several layers of cloth. With the padded side down, place it on the tiles that are set and tap gently with a hammer (Fig. 7-34D), going up and down its length several times. Then, rotate the 2" by 4" piece 90° and repeat the operation. Check both these directions with your level and repeat if necessary to

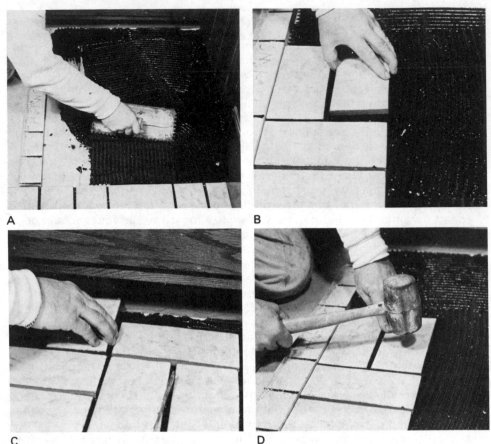

A   B

C   D

Fig. 7-34: Steps in laying a ceramic tile floor.

remove any discrepancies. This tapping action will set the tiles more firmly into the adhesive and will help to achieve a level floor. Allow at least 24 hours for the adhesive to dry before applying grout as described on page 156.

Laying Mosaic Tiles. Mosaic floor tiles, as previously mentioned, come in sheets held together by a web backing. With the small mosaics there is no need to determine a center line. Starting at a corner, measure the remaining distance from the wall to the edge of the sheets. Mark this distance on a sheet of tile. Use a utility knife to cut along the webbing nearest to the marked line. If the tiles have to be cut, carefully mark the cut to be made on the tiles involved. Measure and mark the tiles all around the room. Allow a 1/8" gap between the cut edge and the wall or obstruction. Cut the tiles using the same method as for the wall tile. Finish trimming all the sheets before spreading the adhesive. Mark the floor tile sheets for proper placement, then remove.

Using a notched trowel, apply a floor tile adhesive over a 3' by 3' floor area. Work yourself toward the door. Allow the adhesive to "set up" for 15 minutes, then place small boards on the previously laid tile for you to work from as you lay the remaining tile. Be sure to lay the tile sheets within 30 minutes after applying the adhesive in the designated area. Align the sheets with each other as you lay them (Fig. 7-35A). Allow the same distance between the sheets as the distance between the tiles within the sheets. Press them firmly into the adhesive using a

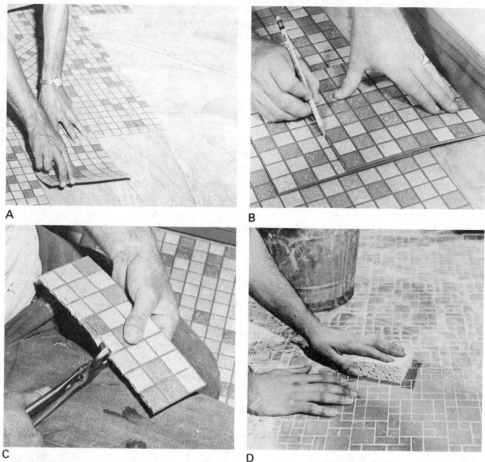

A

B

C

D

Fig. 7-35: Steps in laying mosaic tiles.

rubber surfaced trowel or a padded length of 2" by 4". Adjust the tiles before the adhesive sets up completely. You may have to use a straightedge to level the tops of the tiles. Use a cloth moistened with mineral spirits to wipe off the excessive adhesive which may be forced up between the tiles. If the tops of the tiles are attached to a sheet of protective paper, soak the paper with warm water and a sponge. Peel off the paper and wipe off the excess water. Do not allow the adhesive to set too long before peeling the paper from the surface. In most cases you will need to adjust some of the tiles at this particular time before the adhesive has taken an initial set.

For small floor obstructions, templates are not always needed. Place the sheet face down against the obstruction. Mark the outer edges of the obstruction on the sheet. Pencil in the approximate shape on the webbing, then cut following your outline. Remove any whole tiles which extend beyond the cut edge. Use a pair of tile nippers (Fig. 7-35B and C) for cutting and shaping. Place the cut sheet against the obstruction to check for fit, before pressing the sheet to the adhesive. Individual mosaic tiles can be cut to fill any gaps between the cut sheets and the obstruction. Just cut, scotch tape on the webbing sheet, and lay. Remember that waterproof caulking is required around floor obstructions to prevent water from getting under the tile.

Applying Grout. There are a couple of key points to keep in mind when selecting a grout for your ceramic or similar type floor. First of all, keep in mind that an adhesive installed floor does have a certain amount of movement and "give." Therefore, the usual grouts that are used on walls and made from portland cement set up very rigidly and could crack when someone walks on the floor or through normal movement of the flooring system. Secondly, unlike walls, floors and particularly the grout joints have a tendency to collect dirt. Therefore, you may wish to consider using a darker grout on the floor than you normally would on the walls to reduce this problem. Floors also are susceptible to spills of various types and could be more susceptible to staining than wall joints. Therefore, the darker joints do have a tendency to remain more uniform in color for a longer period of time.

For floor use, there are a number of latex type grouts available that have a tendency to remain more flexible and therefore be more resistant to cracking than the normal portland cement type grouts (Fig. 7-35D). Should your local source of floor tile not have this type of grout available, you may wish to inquire from a local tile distributor, as most of them specialize in this type of application and have these special floor grouts available. These latex grouts usually have a fine sand in them and can best be applied by pouring a small puddle of the grout on the floor surface. Then, using a rubber window squeegee, work the grout diagonally across the joints and into the area between the tiles (Fig. 7-36A). By pulling the squeegee diagonally across the joints you prevent the grout from being "dragged" out of the joint and assure a more uniform and solid filling of these areas between the tile. Before the grout has set up for more than 15 or 20 minutes, you should use coarse rags or burlap to rub the grout from the surface of the tile. Again, be sure you rub diagonally across the joints so you do not rake any of the grout out of these areas. In the event that you did allow the grout to cure too long, you can many times facilitate cleaning it from the surface of the tile with a damp sponge (Fig. 7-36B). In any event, do not wait any longer than 30 to 45 minutes before cleaning the latex grout from the surface of the tile.

A B

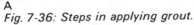
Fig. 7-36: Steps in applying grout.

Allow the grout to set for 24 hours before walking on it. If a haze has developed on the tile from a conventional portland cement grout or if there are still some stubborn smears of the latex grout on the face of the tile, you may find it helpful to add one part of muriatic acid "hydrochloric" to ten parts of water. Wear rubber gloves when using this mixture. Wash the tiles with this solution, then dry them with a clean, dry cloth. Be absolutely certain that you do not bring this solution into contact with your skin or eyes.

Slate Floors

Slate is a natural stone that is either quarried or mined and sawed to the desired dimensions. Gauged slate is then hand-split to the desired thickness (1/4" to 3/8") and ground flat on one surface. The hand-splitting along the natural cleavage on the top surface gives slate its characteristic striated texture. It is also available with both surfaces smooth. Gauged slate comes in a variety of shapes and subdued colors that can be mixed or matched.

Gauged slate can be set in a bed of adhesive over any sound wood (Fig. 7-37), concrete, or composition floor. Loose boards should be nailed. Also, remove any loose paint, grease, or wax. Un-gauged slate (striated on both sides) must be set in a bed of mortar on a concrete slab.

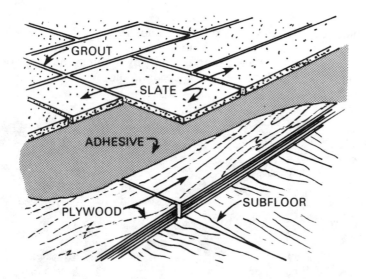

Fig. 7-37: The necessary base for slate.

The setting of slates in adhesive is the same used for ceramic tiles, both in procedure and precautions. But, before actually laying the material, arrange the pieces, without adhesive, in the way you wish them to be in the finished job. Number the pieces in the desired pattern with chalk; this makes them easy to replace in the proper order. The slate can be cut with a hacksaw equipped with a tungsten carbide blade. You can also rent slate cutters from a masonry supply dealer.

Brick Floors

Thanks to real brick veneers and simulated plastic bricks and stones (see page 113), it is possible to have stone and brick floors without special preparation. For example, when a wood subfloor is used with **standard** thickness bricks, the space between floor joists should be reduced by about 25% to compensate for the additional weight. For instance, the rather standard spacing of 16" should be reduced to 12". Before the brick is laid, the base floor or surface should be covered with two or three layers of heavy roofing felt to provide a vapor barrier between the base and the brick. When the brick is to be laid on a concrete or wood base, it must be placed directly on the felt vapor barrier. When either real brick veneer or simulated types are used, they can be set in the adhesive without any special construction details.

BASKET WEAVE   HERRINGBONE   RUNNING

Fig. 7-38: Popular brick patterns: (left) basket weave; (center) herringbone; and (right) running.

Veneer or plastic bricks may be laid in an endless variety of patterns. The basket weave, herringbone, and running bond are the most frequently used patterns (Fig. 7-38), but the actual number is limited only by the imagination. The actual installation (Fig. 7-39) is the same as for ceramic tile.

Fig. 7-39: (Top left) To install bricks, apply a cement (mortar) coat as directed by the manufacturer. (top center) Once the bricks are in place, the grout can be applied by a tube. It can be finished off with a suitable mason's jointing tool or with the back of a metal spoon. (top right) After the grout and mortar have dried, a special sealer is applied. (bottom) The sealer gives the completed floor its beauty.

Normal sweeping or vacuuming and an occasional damp mopping is usually all the maintenance required for brick floors. If desired, interior brick floors may be waxed with a floor wax recommended by a wax manufacturer. Because some floor waxes discolor with age, real brick floors should be sealed with a masonry sealer (a number are on the market) before being waxed. Sealing, even for un-waxed floors, will reduce the likelihood of staining from spilled liquids or clean-ing solutions. Varnish and shellac are not recommended for brick floors because they do not stay in place, and their appearance deteriorates after a short time.

Marble Floors

Marble is limestone that has been more or less crystallized by metamorphosis for centuries. Marble tiles used for floors come in high-gloss and satin finishes, as well as in a range of colors, usually with streaks of other colors running through.

Marble tiles (usually rectangles and squares) suitable for residential flooring can be installed over nearly any subfloor that is sound. They can be set in adhe-sive in the same manner as ceramic tile. After installation, marble floors should be treated with a sealer made specifically for marble and the color of the floor. The floor will need resealing about every six months.

CARPETING

Carpeting many areas of a home, from living room to kitchen and bath, is be-coming more popular as new carpeting materials are developed (Fig. 7-40). The cost, however, may be considerably higher than a finished wood floor, and the life of the carpeting before replacement would be much less than that of the wood floor. Many wise home remodelers will install wood floors even though they expect to carpet the area. The resale value of the home is then retained even if the carpeting is removed. However, the advantage of carpeting in sound absorption and resistance to impact should be considered. If carpeting is to be used, subfloor can consist of 5/8" (minimum) tongued-and-grooved plywood (over 16" joist spacing). The top face of the plywood should be C-plugged grade or better. As already mentioned, construction adhesives are being used to ad-vantage in applying plywood to floor joists. Plywood, particleboard, or other underlayments are also used for a carpet base when installed over a subfloor.

Fig. 7-40: Carpeting looks good in most homes (Courtesy of Armstrong Cork Company).

While a great deal of carpeting is sold on an installed basis, more and more installation is being done by the home craftsman. Carpet materials for the do-it-yourselfer are available in two forms: carpet tiles and roll carpeting. The former, usually in 12" by 12" squares, comes already backed with padding. The carpet tiles are installed in the same manner as resilient floor tiles (see page 140).

Most standard carpeting is available in 12' widths and in continuous rolls. For most do-it-yourself jobs, it is best to use a carpet that is jute, foam rubber, or vinyl foam backed, as these do not require a separate carpet pad. But, when using a vinyl foam backed carpet, check to be sure that the adhesive is compatible with that type of backing; many multipurpose carpet and tile adhesives cannot be used with vinyl foam backed carpeting. Also, jute, foam rubber, and vinyl foam backed carpeting are not recommended for the basement if a water problem exists. In that case, install indoor/outdoor carpet, laid loosely so that it can be picked up and dried if necessary. Tightly woven carpeting is suggested for use in the kitchen, on stairs, and in heavy traffic areas. Shag carpeting, because of its soft, warm appearance, works well in living rooms, bedrooms, or dens.

To estimate the amount of carpeting needed, make a simple sketch on graph paper of the area to be covered. Include all door openings and obstructions. Measure and record these dimensions on your sketch. Remember that the foam rubber applied to the carpet back sometimes results in the carpeting measuring to 1" less than the 12' width in actual finished size. Allow at least 1-1/2" of extra carpet at the ends and sides of the room for wall fitting. If seams cannot be avoided, locate them away from heavy traffic areas. Wherever a seam is to be made, allow the carpet edges to overlap at least 1", since all seams require a new edge cut. Pattern and nap direction should run the same way. This may increase the amount of carpet needed. Mark these measurements and the locations of the seams on your sketch. Then, multiply the length and width measurements to determine the square feet. Divide this total by 9, which will give you the amount of carpeting required in square yards.

Carpet may be installed over any floor that is clean, dry, and free from wax. Waxed floors must be stripped or sanded. Oil base paint should be sanded or scratched. Large cracks or holes in the floor must be filled in. If you are installing carpeting on a basement floor, be sure to seal the floor against dampness with a concrete sealant or binder. Remove furniture from the room. If shoe molding is installed between the baseboard and your present floor, remove it carefully. You may reinstall the molding after laying your carpet.

Loosely lay the carpet over the entire floor before you cut to any walls or do any seaming. Carpet can be cut with a foam cutter or razor knife. Again, the carpeting should be cut at least 1-1/2" longer than your measurements at the ends and sides to allow for wall fitting. Final cutting is completed after the carpeting is placed in its installed position. A straightedge cut is used for the initial placement, 1-1/2" more than the floor area on all sides. If some carpeting has already been installed, the designs should match; check the "pile-lay" to make sure the nap all goes in the same direction. Measure and make chalk lines on the carpet. Place a straightedge on the chalk line and hold it down tightly. Using a foam cutter or razor knife, firmly and evenly cut along the straightedge. Mark the location of the obstruction on the carpet. Chalk a line from the edge of the carpet to the obstruction. Place a straightedge on the line and make the cut. Hold the carpet tightly against the obstruction. Cut and fit the carpeting around the obstruction.

Roll carpeting can be laid in three ways: (1) using pressure-sensitive, double-faced tape; (2) employing wooden tackless stripping; and (3) using adhesive. In this book, we will concern ourselves with the latter. The installation procedure is as follows: Loosely lay the carpet on the floor and check that it fits the room, then roll it up. Align the carpet with the wall, trowel the carpet adhesive over the floor along the wall as a starting strip, then press the carpet into the adhesive. Trowel more adhesive onto the floor and continue to unroll the carpet. Brush out any air bubbles that may appear. Do not stretch the carpet or in any way allow it to adhere under tension. Trim the carpet along the walls. When cutting any extra carpet from the edges, press the carpeting tightly into the corners where the floor and wall meet. Cut in toward the wall rather than into the floor to insure a snug fit. To complete the job, install metal edging at doorways and shoe molding along the walls.

When making seams, lay the first piece of carpet on the floor where the seam is to be made, but do not secure the carpet in place. Then, overlap the pieces at least 1". If the carpet has a definite design, the pieces must overlap until the designs match. On the upper piece of carpet (Fig. 7-41), make a mark 1" from the edge on each end (A). Snap a chalk line on the carpet between the markings (B). Follow the same procedure to mark the edge of the bottom piece. Align the edge of the upper carpet with the chalk line on the bottom piece (C). Using the top chalk line as a guide, place a straightedge on the upper piece. The cut should be made 1/2" from the edge of the upper carpet. Cut through both pieces with a carpet cutter or razor knife, on a slight bevel inward toward the carpet. You now have a clean, matching seam. Then, snap a chalk line on the floor where this seam will fall. Align both pieces with the chalk line. Fold back both pieces of carpet half-way and spread floor adhesive under the first piece. Re-lay the first piece of carpet onto the adhesive area, following the laying instructions previously given.

Trim the second piece, checking for a snug fit by pushing it slightly over (1/8") the first piece. Apply the adhesive at the edge of the backing on the first piece of carpet. Do not apply to or allow the adhesive to touch the surface yarn. Re-lay the second piece (overlay 1/8") onto the adhesive area. Form the seam by hand tucking or crimping in the overlapping piece. Clean the seam with a damp cloth and gentle rubbing. Continue to lay the rest of the carpet.

Carpeting Stairs. Installing carpet on stairs will improve their appearance and muffle sound. Left over pieces from carpeting the room can be trimmed to fit the stairs. Coat the cut edges of your carpet with latex adhesive to prevent fraying.

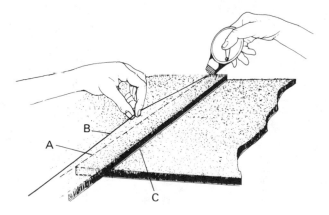

Fig. 7-41: Making a seam.

Runners for stairways are also available that come in standard widths of 18", 22-1/2", 27", and 36" and are sold by the lineal yard. To find the amount of carpeting needed, measure in inches the depth of one tread and the height of one riser; add the two measurements together and multiply by the number of stairs; to this figure, add the length of any landings and divide the resulting amount by 36 to determine the number of lineal yards.

When employing carpet adhesive, place it on each tread and riser, as well as tread nosing. Start at the top landing; align the carpet with the stairs. If the top landing requires carpeting, first apply the adhesive to it; then, using carpet tacks, tack the top edge of the carpet around the landing (Fig. 7-42A). Do not tack the carpeting down on the stair edge; rather, tack it (Fig. 7-42B) on the riser (Point **a**), directly under the tread (Point **b**).

When pulling the carpet down over the nose of the steps, smooth it firmly into place. Tack it into the riser, under the stair tread. Work slowly and carefully, patting the carpet into place. Be sure there are no bulges or air pockets. As you proceed downward, press the carpet into the crease of the stairs (Fig. 7-42C) by hammering a wedge-shaped piece of wood into the area. Secure by tacking along the crease. Always remember that stair carpeting should be laid with the pile facing down the stairs for maximum wear resistance. Check the sweep of the pile by stroking it back and forth lengthwise. The smoother stroke is called the "lay" of the pile direction. At the bottom of the steps, trim the carpet and tack it securely into place along the bottom edge.

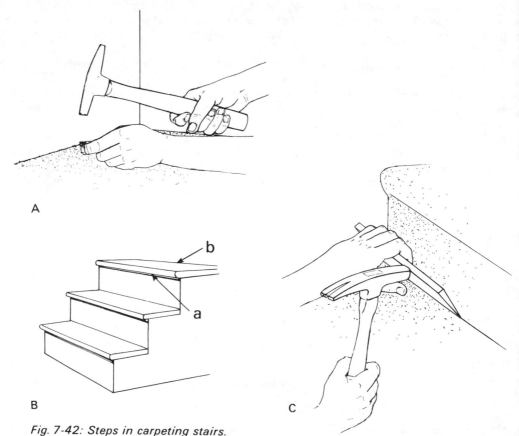

A

B

C

Fig. 7-42: Steps in carpeting stairs.

Specialty Adhesives and 8
How to Use Them

There are a great number of adhesives, glues, and cements on the market today, and many of them could be classed as specialty types. In the previous chapters, we have concerned ourselves mainly with wood glues, contact cements, and construction adhesives. In this chapter, we take a look at the adhesives with special characteristics that make them best suited for a particular type of bonding job. But before taking this look, it is important to keep in mind that the how-to-use instructions given are only typical for that class of adhesive and should only be used as a general guide. Always follow the instructions that come with the brand you buy. These instructions should state the type of materials that can be bonded, the length of drying time, the solvent or cleaning agent, application, and whether or not pressure is required on the pieces being joined. All labels state very clearly any hazard which may be encountered and every precaution that should be taken. In other words, if you want to obtain the best possible results, these instructions should be observed to the letter.

Cyanoacrylate Adhesives

In many ways, this is a remarkable adhesive. They are sometimes referred to as "super glues" and provide tremendous strength and a quick setting time—10 to 30 seconds. A group of chemicals—cyanogen and acrylic resins—when combined, are anaerobic, meaning that they cure in the absence of oxygen. Most cyanoacrylate manufacturers usually go to great lengths on their packages to point out that the container is purposely not filled to its capacity. If it were, it would preclude any oxygen in the container and, as a result, the contents would cure and could not be used. Therefore, space must be left in the container for the oxygen.

Since cyanoacrylates are anaerobic, they can be used only on solid, *nonporous* materials, adhered either to themselves or to another nonporous surface. This includes metal, china, glass, jewelry, rubber, glassware, and most plastics, ceramics, and leather. Cyanoacrylates do not work on soft or absorbent surfaces such as fabrics, paper, or wood. They do resist temperatures up to about 160°F, continuous immersion in water, and many chemicals.

Cyanoacrylates, despite their great publicity, should not be used with abandon. The glue will bond skin together almost instantaneously, and there have been cases of people having to have surgery to get fingers unglued. Also, it is not economical to use a cyanoacrylate if any other type of adhesive can do the job instead. Cyanoacrylates are the most expensive adhesives sold on the consumer market.

Instructions for applying a typical cyanoacrylate are as follows:

1. Remove all dust, oil, or grease from the surfaces to be bonded. (This is a general rule for all specialty adhesives.)

2. Remove the outer cap; flick the end of the tube with your finger so that there is no liquid at the tip which may spout when opened.

3. Hold the tube at the neck. Point the metal tube away from the eyes, and insert a pin into the nozzle, far enough to pierce the metal tube. Never use a scissors to enlarge the hole.

4. Squeeze the tube *gently* and apply to one surface only. One drop per square inch is sufficient since excess adhesive gives a slower setting time and sometimes a weaker bond.

5. Place the second surface in contact with the surface having the adhesive and quickly rub the two surfaces together with light pressure to spread the glue. Parts must mate; adhesive will not fill gaps. *Caution:* Do not get glue on your fingers as it will bond on contact. *Use care in handling.*

6. Press parts together for 10 seconds, or until the adhesive sets. Allow it to completely harden for several minutes before use. Full strength is attained in 3 to 6 hours.

7. Wipe excessive adhesive from the dispenser tip with a cloth dipped in nail polish remover or acetone, replace the outer cap, and store in a cool, dry location with the tube in an upright position.

Epoxy Adhesives

Originally developed for general industrial use and especially for such stringent uses as the construction of aircraft components, epoxy adhesives can be used to bond hard-to-stick materials, such as glass, metal, concrete, porcelain, and many plastics. Hardening is by chemical (catalytic) action rather than by drying or evaporation of solvents so they will harden even inside an airtight container or under water.

While the resin portion of an epoxy system has certain inherent characteristics relative to adhesion, the performance of a given formulation in its cured environment depends a great deal on the type of catalyst or hardener that the manufacturer has selected to provide with the epoxy resin. Those generally used in the consumer markets are of the polyamide resin type. They are easier to use than other types of curing agents. For one thing, the proportion of epoxy and curing agent to be mixed is not real critical. You can guess at an approximate 50/50 mix by volume, and even if you are not exact, you will generally acquire a good cure. By increasing the hardener you can impart more flexibility into the cured bond. Generally, these proportions can be altered by as much as 60% hardener and 40% resin and vice versa. As you increase the epoxy resin portion however, you might acquire greater strengths, but you lose some of the impact resistance as the cured bond becomes somewhat brittle.

Clear, as well as colored (including white and metallic), epoxies are available. The strengths developed are quite high, up in the thousands of pounds per square inch shear, again depending on the type of hardener used and the proportion of mix. They have excellent resistance to solvents and most household chemicals, and while they tend to be water-resistant, they should not be used as a waterproof glue. The temperature extremes under which they will perform are wider than for most consumer-type glues and adhesives. However, here again there are limitations. Continuous temperature cycling from hot to cold has a very adverse effect on most adhesive and glue bonds, and the epoxies are no exception, even though they will normally withstand more than most other materials. Continuous operation at temperatures above 200°F could create problems.

When used within the guidelines outlined above, epoxies are a very versatile group of products. They can be used to repair leaky pipes, drain lines, radiators, engine blocks, and so on. They are strong enough to handle such jobs as installing fixtures on ceramic tile and other wall surfaces and permanently repairing wrought iron railings and furniture. They make excellent wood joints; however, their cost is generally prohibitive for this type of application.

All epoxy formulations found in the consumer market are packaged in two separate containers—the epoxy resin portion and the catalyst or hardener portion. Where small amounts are needed, squeeze out equal parts of resin and hardener on a clean surface. If dispensing from a larger container, such as a can or jar, never use the same tool to "dip" from each of the containers. This will contaminate either the resin or hardener and will ruin them for further use.

Some formulations are made purposely thick. These particularly have to be dipped rather than poured together (Fig. 8-1). Thoroughly mix the resin and hardener to a uniform color. Once they are mixed, they immediately begin to cure, so you have a limited pot life. It is a good practice to never mix more resin and hardener than you can use in 30 minutes.

Once the typical polyamide-cured epoxy adhesive is properly mixed, apply a thin coat to the surfaces to be joined. Because of their high bonding strength and the fact that they will not shrink or change dimension as they dry, no clamping pressure is required, though parts should be tied or taped together to hold them in place while the adhesive sets. Allow the epoxy adhesive to set for 4 to 6 hours, or overnight for full strength. For faster drying, apply heat from a device such as a heat lamp—never an open flame. At 250°F, the drying time is less than one hour (Fig. 8-2).

Water-phase type epoxy may be used either as a coating or as an adhesive. It is used primarily for bonding fiberglass over boat decks or porch seams. Once hard, they are highly water-resistant. It is a two-part type; the two liquid components are mixed in equal proportions or as directed by the manufacturer. Complete setting time is approximately 10 hours. It is made in clear, white, and colored forms and can be tinted with the same dry colors used in cement work. Unlike other epoxies, the water-phase types can be washed from brushes and tools with water. Polyamide- and amine-cured types are best applied with cheap brushes, which are then discarded.

A recent development in epoxy cements is the hand moldable stick. This two-part epoxy handles like putty, but hardens like steel. Like other epoxy cements, it has outstanding holding and sealing power. To mix, unwrap the sticks suffi-

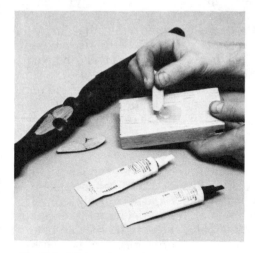

Fig. 8-1: Mixing the resin and hardener to produce an epoxy adhesive.

Fig. 8-2: Using a heat lamp to increase the curing time of an epoxy adhesive.

ciently to cut equal lengths from each bar. Roll, knead, and press the lengths together with your hands to form a ball of uniform color. For easier application and less sticking of compound to your fingers, wash and dry your hands after mixing. Then, work the epoxy firmly into the repair area. Smooth it with wet fingers or a tool, if desired. Remove any excess before hardening occurs. Finish may be applied in 12 hours. *Caution:* The epoxy may irritate sensitive skin; wash your hands with soap and water. Close the overwrap on each unused length of the sticks to prevent moisture absorption and hardening.

Polyester Adhesives

While polyester resin adhesives will bond to a variety of materials, they are most widely employed in building fiberglass boats and, like water-phase type epoxies, for bonding fiberglass fabrics to boat hulls and decks. They are also used in stonework and marble repairs. They can be used clear or colored with pigments sold for the purpose.

A polyester resin adhesive is a two-part type. A liquid resin and a liquid catalyst are mixed just before use. Typically, a specified number of drops of catalyst from a small squeeze bottle are stirred into a quart or pint can of the resin. A portion of the resin may be poured off and mixed with a corresponding portion of catalyst. The amount of catalyst is important; an excess will harden the resin before it can be applied.

Temperature of the working area is also a factor, because hardening time shortens as the working area temperature rises. Temperature-hardening time charts are available from suppliers for many resin brands. Do not apply a chart for one brand to another brand of resin; formulas vary widely.

Polyester resin glue is sold in cans, usually with a plastic tube of catalyst in a plasic overcap, by boatyards, marine suppliers, and large hardware outlets. The shelf life of unused resin varies with the brand and storage conditions, so buy it fresh, as needed.

Polyvinyl Chloride (PVC) Adhesive

Polyvinyl chloride bonds glass, china, porcelain, marble, metal, and many hard plastics. It can be used on wood and other porous materials as well. It is unaffected by alcohol, oil, or gasoline.

When using a polyvinyl chloride to bond nonporous materials, apply the adhesive evenly to one surface—be sure it is clean and dry—and wait 2 to 3 minutes for the spread cement to become tacky. Then, press it against the other part to be bonded; no clamping pressure need be applied as long as the parts remain in contact and in position. On porous surfaces, the adhesive is applied to both surfaces and allowed to dry to the touch. Then, the adhesive is reapplied to both parts and they are pressed together. When applying PVC to a porous and nonporous surface, apply the adhesive only to the nonporous surface. The adhesive dries in 10 to 30 minutes. Any squeeze-out can usually be removed with either acetone or lacquer thinner.

Cellulose Cements

Cellulose nitrate or nitrocellulose cements, often called household cements or model airplane cements, are probably the oldest of the so-called "specialty adhesives" still on the market. They usually dry clear and set quickly (from 2 to 10 minutes), with some pressure put on the joint, but the material takes about 24 hours to cure completely. However, most celluloses are flammable before they are dry. Their fumes are toxic and dangerous; therefore, be sure to work in a well-ventilated room. Do not work with cellulose glues near a flame (including stove pilot lights or lighted cigarettes).

A cellulose cement is good for wood, metal, leather, paper, glass, and ceramics. It is also fine for jewelry repairs. But, these cements should not be used on objects that will be put under stress. Most brands are moderately water-resistant, but they all will weaken if the cemented joint is soaked for any length of time. Because of the relatively low solids content of this type of adhesive, there may be slight shrinkage when dry. That is, the drying action tends to draw the parts of a joint together with a fine, sometimes almost invisible glue line. This can be most advantageous in model making or fine woodworking.

When applying a typical cellulose cement, spread a thin coat on each surface to be joined. Then, press the surfaces together. Clamping is generally not required, especially if a second coat of cement is applied after the first coat has become slightly tacky. Press the two parts together and hold them under hand pressure for 2 or 3 minutes. Bonding nonporous materials usually takes a little longer. Do not spill the cement on painted surfaces or fabrics. Use acetone or ethyl acetate to remove any spilled or excess cement.

China/Glass Cements

In recent years, cellulose cements have lost their general-purpose popularity to aliphatics, epoxies, and cyanoacrylates. They are also losing out in the china/glass repair field to the newer specialty cements. These china/glass cements, made under several different formulations, are fast-setting and powerful, and they dry crystal-clear like the cellulose types. In addition, they will withstand detergents and a dishwasher's superheated water. Further, they are nonflammable and contain no harmful fumes.

Before applying any china/glass cement, make sure that the surfaces are clean, dry, and free from oil, soap, wax, paint, varnish, previously used cement, and any other residue. Apply the cement to one surface only. Then, position the pieces as closely as possible without touching. When they are properly aligned, press them together for about one minute. Some cement may squeeze out at this point; with a clean cloth, wipe away the excess while it is still wet. Wait for the cement to become clear (2 to 5 minutes), if applying additional pieces. If so, handle with care, avoiding any stress. Let the repaired china or glass item stand overnight before using it in a dishwasher. Most china/glass cements will *not* withstand high oven heat.

Acrylonitrile Adhesives

Acrylonitrile, or buna-N adhesives, as they were once known, were originally developed for aircraft use. They consist of a one-part, thick beige liquid that will hold almost any material with a completely flexible and waterproof bond. They can be used to fasten fabric to metal or wood as well as metal to glass. They can even be employed for bonding patches to work clothes or for repairing rips and tears in carpets, tents, and sails, with greater strength than when mending with thread. While it can be used to bond other materials to wood, it is not recommended as a wood adhesive.

There are several ways of applying acrylonitriles. In the wet bonding method, the adhesive is applied to both joining surfaces, allowed to dry to a tacky condition, and then the parts are joined under sufficient pressure to maintain good contact. This method of fastening can be used for most materials.

In some fabric-to-fabric bonding operations, the adhesive is applied to the surfaces to be held, allowed to dry completely, and then set under pressure by heating from the outside of one fabric piece, as with a flatiron. The heat required to achieve this bonding procedure may range anywhere from 175 to 325°F.

Acrylonitrile adhesives can also be applied by the reactivation method. In this application, the adhesive is allowed to dry completely as before, but, to make the bond, one coated surface is reactivated by wetting it with a solvent solution such as acetone, ethyl acetate, or methyl ethyl ketone. The parts can then be joined together with hand pressure. This method of joining is generally used for larger pieces of work. Generally speaking, this is a highly specialized method of application used primarily in industrial applications and is not recommended for home or do-it-yourself use.

Fabric Mending Cements

These special adhesive formulations eliminate the need for sewing and darning torn fabrics. They enable you to repair tears and mend holes in wool, cotton, canvas, leather, denim, and most other fabrics. Use them with or without a supporting patch to repair canvas awnings, luggage, sportswear, tents, and clothing. When properly applied and thoroughly dried, the patch will withstand normal washing and ironing, though some brands are susceptible to attack by dry cleaning fluids. But, before attempting to make a mend, be sure to test the cement for bleed-through in an inconspicuous spot on the fabric.

To properly apply any of the mender cements, the fabric should be clean, dry, and free from oil, grease, wax, or dirt. Then, at room temperature, spread a small amount of cement on both surfaces and join with light pressure. Allow it to set 24 hours before using. If you wish the fabric article to withstand laundering, it must be ironed for a minute at the wash-and-wear setting (300° F) of the iron, *after* the cement has dried 2 hours. Allow it to cool before using.

To clean up any excess cement while it is wet, use water. When the cement is dry, remove the excess with alcohol; but, first test for fabric compatibility in an inconspicuous area.

Latex Adhesives

Many of today's fabric mending cements are still latex-based adhesives. Being an emulsion system, these adhesives cure or dry by water loss and form a flexible, rubber-like bond that is resistant to repeated washing. In most cases, latex adhesives are applied in the same way as the mending cements described above.

Plastic Mending Cements

With the increased use of plastic items, it was inevitable that the adhesive industry would produce a glue for their repair. There are now on the market cements that are designed especially for repairing flexible and rigid plastic objects, including such materials as styrene, vinyl, acrylic, and phenolic resins. These clear, flexible adhesives can be used to mend torn plastic raincoats, plastic toys, above-ground swimming pools, and similar materials. They are also ideal for assembling plastic models of all kinds, as well as for cementing patches onto inflatable toys and beach equipment. Several brands can even be used on wet surfaces and under water, so they will permit you to patch plastic swimming pools without having to drain them.

Most flexible items require a patch from the same or similar material as that being repaired. The patch should be large enough so that it extends about 1/2" beyond the edges of the mend. Before applying the adhesive, lightly sand or roughen the surface of the patch and the area where it is to be placed in order to improve adhesion. Then, apply the cement all around the edge of the underside of the patch. Press the patch into position and maintain pressure with clamps for 2 or 3 hours, or until dry. Allow the mend to dry overnight before using the item. For a less visible repair, the patch may be applied to the underside of the item being mended.

To make a rigid item repair, just fill the crack with cement and allow it to dry overnight before using. Most plastic mending cements can be cleaned up with nail polish remover or acetone.

Plastic Rubber Cements

Although basically a sealant and waterproofing compound, plastic rubber cement is also a strong adhesive which repairs many kinds of rubber objects permanently. Consisting of neoprene rubber in a creamy or paste-like consistency, it comes in a translucent amber color, as well as in black and white. It is used for repairing rubber boots, for mending leaks in rubber rainwear, and for such odd jobs as making your own insulated handles on all kinds of tools. It sticks to almost any clean, dry surface—porous or not—therefore, you can also use it for bonding all sorts of dissimilar materials. It will adhere to glass, metal, most plastics, leather, and fabric. You can also use it to seal and caulk seams on wood, metal, and canvas boats, as well as for permanently insulating and weatherproofing electrical connections and exposed wires.

When using plastic rubber cements for repair work, they are applied in the same manner as plastic mending adhesives. That is, they are used with a patch on flexible surfaces and as a crack filler on rigid items (Fig. 8-3). Where adhesion is not wanted, sprinkle the surface with talcum powder.

Plastic Metal Cements

There are a number of plastic metal cement formulations on the market. Most, like the plastic aluminum, for example, which contains atomized aluminum and vinyl resin, are available in either a liquid or putty form. They come to you ready-mixed, ready to use as a filler or adhesive. They adhere to practically any surface, and once dry and hard, they can be burnished with any metal to give the surface a metallic appearance. Most cements, when dry, can be filed, drilled, tapped, or sanded, and may be painted over with enamels or lacquers.

When applying a typical putty-type plastic metal cement as a filler, proceed as follows:

1. Be certain that the surface to be repaired is clean and free from dirt and grease. Roughening the surface slightly with abrasive paper will insure much better adhesion.

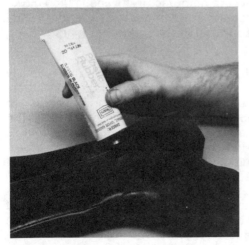

Fig. 8-3: Applying plastic rubber cement to a rigid surface.

Fig. 8-4: Adding concrete bonder to a concrete mix.

2. Apply the cement as it comes from the tube, leveling it out with a knife or spatula, if necessary.

3. It dries to metal-like hardness in 3 to 4 hours at normal room temperatures. For filling deep holes, etc., apply in layers of 1/8", allowing each application to dry thoroughly before adding another. For thinning or clean-up, use acetone or nail polish remover.

To use a typical liquid plastic metal cement as an adhesive, apply a thin coat to both sides of the area being repaired. Then, press together firmly. Most cements will start to bond immediately; however, allow it to dry overnight. Some metal cements are nonmetallic in content and thus can be used to repair enamelware pots and pans.

Rubber-Based Adhesives

A rubber-based or styrene-butadiene adhesive is a one-part water-resistant type that bonds to most nonporous materials, including metal, masonry, glass, and many plastics. It can be used to replace loose tiles and bricks, to attach bathroom and kitchen fixtures to walls (tile and non-tile), and to bond letters onto nameplates. It can also be employed to hold nuts on bolts in place of lockwashers and has excellent gap-filling properties that provide a grip on irregular surfaces. A rubber-based adhesive can be drilled, tapped, sanded, and painted when hard. It can even be applied as a sealer under water for emergency repairs. It should not be used where exposed to gasoline or oil, however, as both act on it as solvents.

Before applying a typical rubber-based adhesive, be sure that both surfaces are clean and tight. Then, butter the adhesive on one surface and allow it to air-dry a few minutes until it becomes tacky. Once the two surfaces are joined, no clamping is usually required. Any excess adhesive can be removed with mineral spirits or turpentine.

Polysulfide Adhesives

While polysulfides are better known as a marine caulking material, they are also a strong, completely waterproof adhesive with great flexibility and unusual vibration-damping qualities. The two-part form requires the mixing of two thick pastes. The one-part form sets by reaction with moisture in the air. When set, polysulfide is actually synthetic rubber.

Since polysulfide formulas vary, follow the manufacturer's application directions and plan on the setting times indicated. Polysulfides are generally industrial-type products and are not too often found in the consumer markets.

Silicone Adhesives

Silicone rubber is also primarily a sealant and caulking compound that is widely used on a variety of construction surfaces, particularly in tub and shower areas. First developed for sealing space capsule components, this material is packaged in squeeze tubes and dries to a permanently flexible rubbery consistency that will neither crack nor dry out. It withstands all ordinary chemicals, as well as extremes of heat and cold. It is useful indoors or out.

There are several types of silicone adhesives on the market, each with their own interesting characteristics. For instance, the typical silicone glass/ceramic adhesive makes a strong, tight seal that stands up to not only soapy suds, but also oven heat and freezer cold. It is safe for contact with food, and it will not shrink or crack. But, when using a silicone to repair small contact areas, such as cup handles or plate edges, hold the pieces together with tape until the adhesive cures. It does not have the quick setting time that other glass/ceramic adhesives do.

Silicone adhesives are generally used on glass, ceramic, metal, china, and many plastic surfaces; some are suitable for wood, canvas, and rubber. Because of the characteristics of the different types of silicone, it is a *must* to check the manufacturer's instructions before applying.

Concrete Bonders

Concrete bonder, concrete adhesive, or concrete patch, as it is sometimes called, is a highly water-resistant acrylic resin formulation designed to serve five basic applications. There are two basic types generally found. One is an acrylic system and the other a polyvinyl acetate system. The acrylics are much more waterproof and should be used on all exterior applications.

1. **Adhesive Tie Coat.** A good concrete bonder makes it possible to bond new wet stucco, plaster, or cement mixes to old concrete, stone, wood, metal, and other similar surfaces. Before applying, be certain that the surfaces are clean and free from dust, dirt, and loose materials. It is not, however, necessary to do any keying or joint raking. Spread the concrete bonder—using a brush, spray gun, or roller—over the entire surface to be repaired in much the same manner as you would paint. Do not, however, thin or dilute the adhesive, unless so stated by the manufacturer. (Some concrete bonders are sold in concentrated form and can be diluted with water to provide a higher volume of bonder.) The new concrete mix, plaster, or stucco may be applied over the concrete bonder for a period of up to 2 hours.

2. **Fortifier.** It imparts extra strength and toughness to portland cement and dry-ready mixes. Thoroughly mix all the dry ingredients of the concrete mix; then, add the concrete bonder (Fig. 8-4). Various manufacturers provide different concentrations of liquid. Therefore, be absolutely certain to follow the manufacturer's directions for mixing methods and ratios in order to derive the best results.

After the bonder is mixed, the water can be added to the concrete mix to obtain the desired consistency. If greater concrete strength is desired, additional bonder can be added to the mix.

3. **Underlayment.** Concrete bonder can be used to level uneven floors of concrete, wood, old tile, or other surfaces prior to the installation of new flooring materials. To use the bonder for this application, mix it in the same proportions as you would for using it as a fortifier. Apply the mix to the old surface with a trowel and then smooth it to a level finish. The mix may be troweled from a featheredge to a thickness of 2". All cracks and voids in the subfloor surface can also be filled and troweled smooth.

4. **Primer/Sealer.** It can be applied to a porous material such as wallboard, wallpaper, plaster, concrete, and stucco prior to painting or grouting. When used for this purpose, the bonder is usually diluted approximately 2 parts to 1 with water and applied evenly by brush, roller, or spray gun. Allow it to dry for 45 minutes or more before applying a water-thinnable mix. The surface must be *thoroughly* dry before putting on paint.

5. **Decorative Nonslip Coating.** Concrete bonder can be used as a nonslip coating around pools, walks, steps, boats, boat docks, and on metal, wood, or old concrete. Using a concrete ready-mix and bonder mixture in about the same proportions as described for a fortifier, mark off the area coated with masking tape, and with a trowel, spread a heavy coating (1/16" to 1/8" thick). Then, using a stiff bristle brush or broom, brush the surface to make a rough texture. Acrylic paint colors may be added to the wet mix to provide color to the coating.

Hot Melt Glues

Hot melt glues—made of either thermoplastic polyamide or polyethylene ad-

Fig. 8-5: Using an electric glue gun.   Fig. 8-6: Applying rubber cement to paper.

hesive bases—are designed to be used in an electric glue gun. They come in a solid stick form, usually about 1/2″ in diameter and 2 and 4″ long. The glue stick is inserted into the back of the gun and extruded through the nozzle by either pushing the stick with your thumb (Fig. 8-5) or pulling the trigger, depending on the tool. Good for bonding most porous materials, they are nontoxic, odorless, nonflammable, and not affected by moisture. However, you must work fast with hot melts since most have an open time of 15 to 20 seconds and set in 60 seconds.

While various glue guns have different features—preset thermostat, aluminum melting chamber, self-standing design, two-cartridge storage, and so on—the following basic instructions hold for all models.

1. Insert the glue stick cartridge in the gun, in the proper chamber as detailed by the manufacturer's instructions. Some guns require the insertion of two cartridges when first used, because they feature a reserve glue compartment.

2. Plug the gun into a 120-volt electric circuit (unless it is designated as a 240-volt unit), and allow 3 minutes for warm-up. Take care not to touch the heating chamber. The temperature in this area may reach as high as 400°F.

3. When the 3-minute warm-up period has passed, squeeze the trigger of the tool to dispense the glue. From the time you start applying the adhesive to the surfaces to be bonded together, you have about 10 seconds of time. Work quickly. Do not try to cover a large area. Place the two surfaces to be bonded together within 10 seconds. As with any adhesive, the two bonding surfaces must be clean.

4. After 60 seconds, the adhesive will attain 90 percent of its bonding ability, which is sufficient to hold the surfaces together.

5. When you are finished with the glue gun, pull the line plug and rest the tool on a nonflammable material or hang it on a hook to cool. Allow the glue to remain in the tool; it can be reheated repeatedly without affecting its bonding ability. Never change the nozzle when it is hot.

Paper Adhesives

Although many glues work on paper, some rubber cements (Fig. 8-6), mucilages, library pastes, and animal or vegetable glues are made solely for paper bonding. They are handy in the home, office, or school.

Most paper adhesives are applied with a brush to one or both surfaces (a stronger bond) and then joined by pressing the two paper surfaces together. Let it dry before moving the glued paper.

There are other specialty glues and adhesives that are available. For instance, there are paper and vinyl wall covering adhesives that are employed to hang wall covering on walls and ceilings. For woodworking, there are special sanding disk cements used to install sandpaper on the pads of orbital sanders. Because the list could go on and on, it is impossible to name all of the specialty adhesives in one book—even several. The most common ones—filling the needs of most home uses—are given in this chapter. If you cannot find the one you want, check your local hardware, lumber yard, home center, or marine supply dealer for his advice. They usually have the answers.

Why Didn't It Stick? 9

You have spent a great deal of time and money accumulating and purchasing beautiful materials for your project. You have spent hours planning and working on it only to find that "the adhesive just did *not* stick." Why?

The first reaction you might have upon discovering that your adhesive or glue did not hold would be to think something is wrong with the product itself. Although it is possible that you obtained a "bad batch" of adhesive or glue, it is really very unlikely. With everything in its proper perspective, you can assume that the manufacturer produces these products in large quantities. In addition, most reputable manufacturers include a quality control program in the production of their product. Therefore, if there was something wrong with the adhesive or glue, the manufacturer probably would have found it and would have recalled the product. It is also unlikely that the small quantity you received was the only portion of the thousands of gallons produced that was bad. It is much more probable that there is another cause for your problem.

In fact, the answer to the "why it will not stick" problem usually comes only after weeding through all the possible causes and then determining the solution that will help you remedy the condition or avoid the trouble in the future. It is necessary to collect all possible facts related to the trouble. The list of potential trouble areas to check could become somewhat lengthy. As described below, there are a few that seem to occur more frequently than others and should always be checked first.

Freeze/Thaw Stability

Some adhesive products, particularly those in the latex classification and the general emulsion class, are subject to freezing. While a good portion of these are called freeze/thaw stable, others can be harmed by freezing. The signal to you is the notation on the label "Keep From Freezing." The good part of this is that you usually do not have to worry about using a product that has been damaged by freezing. If it is not freeze/thaw stable, the adhesive coagulates upon freezing, and you will not be able to get it out of the container. Those that are freeze/thaw stable can generally go through a number of cycles of freezing and thawing without damage.

Improper Storage of Product

It is always a good idea to store any adhesive at room temperature for 24 to 48 hours prior to use during the colder winter months. Even those that are not harmed by freezing have a tendency to "thicken" when cold and are hard to use.

Some adhesives are packaged in fiber containers such as cartridges. These can be damaged and weakened if stored in a damp area, and the adhesive may lose solvent and become heavy or even hard. Sometimes, this is undetectable until you try to use the adhesive.

Incorrect Product for Application

How many times have you had an adhesive or gluing task come up suddenly and you began to rummage through the workshop for a left-over container of adhesive? Sometimes you were able to find the proper materials and at others you were not. Hopefully, you are now much more aware of the nature of glues and adhesives than you were before and you will remember that not just any glue or adhesive will work on any application. While the left-over adhesive that you found might work, chances are that it is probably too old to trust by now. It is also possible that it might not have been properly sealed and stored the last time that it was used.

Stop and think for a moment before you use just anything that you have available. You might destroy or ruin one of the substrates that you are trying to bond and will then have to spend a lot of extra money buying new materials. Always make certain that the adhesive you have purchased was designed and formulated to satisfactorily perform on the project you have undertaken. There have been occasions when an untrained or unknowledgeable clerk has made an incorrect recommendation of an adhesive for a given task and the result was most unfortunate. If the label makes no mention of your application, it would be wise to ask about the uses of the adhesive or to look for one that is appropriate.

Adverse Effects of Adhesive on Adherend

It is important to be sure that the materials being bonded and the adhesive or glue being used are compatible. For instance, polystyrene foams, commonly bonded to many construction surfaces, are literally dissolved when brought into contact with many of the solvents used in the formulation of panel adhesives. Also, when applying an adhesive to a decorative surface, it often has a tendency to bleed through the surface. Yet another common problem with adhesives concerns their use on plastics. The solvents used in various adhesive formulations may harm the plastics and cause them to be unreceptive to bonding.

If you are in doubt about the effects that an adhesive or glue might have upon the materials being bonded, try some in an inconspicuous spot or on a scrap piece of the material. This might save you a good deal of time, money, and grief. Bond together scraps of the materials that you desire to adhere. After 24 hours, pry them apart and examine the glue line. If a good bond was not attained or if either surface was damaged, then you are not using the correct product.

Materials Unreceptive to Bonding

Unclean Surfaces. One of the most common causes of failure on adhesive installations is an unclean surface on the materials being bonded. Remember, if you are bonding to dust, dirt, loose scaly paint, oil, grease, etc., the bond that you get will be no better than the adhesion of these foreign materials to the substrates. Always thoroughly clean and degrease any surfaces to which you are bonding. If you use solvents of any type for cleaning purposes, be careful and observe all recommended safety precautions. Never use gasoline or fuel oils as they leave an oily residue that is difficult to bond to. A reasonably safe general purpose solvent would be mineral spirits; and on small areas, lighter fluid might work.

It is sometimes desirable to clean walls, ceilings, and floors with a strong detergent or something such as trisodium phosphate. On floors, all waxes must be removed from old existing floors before you attempt to adhere any material to the surface. If there is loose tile or flooring of any type, it should be removed and the area or depression leveled with a latex/concrete bonder or flooring underlayment compound.

Tolerances Too Great (Too Much Gap Filling). Another common cause of failure on adhesive installations would be gaps or dips in the subfloor. Even smooth troweled concrete that is clean and dry can sometimes cause problems through "puddling." For example, if you were to hose down a concrete floor, even a new one, with water, you would find some puddles of water accumulating in the low areas. Should you be putting down a rigid flooring material, such as slate, quarry tile, or wood parquet, it is possible that a gap would form between the flooring and the subfloor, as the flooring would tend to bridge over the low points. In severe cases, the adhesive that was troweled onto the subfloor may be prevented from coming into contact with the flooring and an inadequate bond would be developed.

These low places in floors as well as all cracks should be leveled and filled with a suitable latex/concrete underlayment before any attempt is made to install flooring. Mastic type adhesives will perform some gap filling functions, but all too often so great a burden is placed on them that a poor installation will result.

Surfaces Not Dry. One final problem which may cause materials to be unreceptive to bonding would be a surface which is not completely dry before adhesive is applied to it. Most adhesives are not designed to bond to wet surfaces, and you can run into a lot of trouble if the substrates are wet or, in some cases, even damp. You should be particularly cautious of damp masonry walls and concrete slabs and eliminate the source of moisture before attempting to bond anything. Even though a wall or floor may appear dry, surface water can seep through masonry walls during heavy rains and hydrostatic pressures can develop in floors. Adhesives and glues are not waterproofing compounds and should not be used for this purpose even if they are called "rubber" based, waterproof, or water-resistant.

Improper Application of Adhesive or Glue

Incorrect Amount. All too often, a fortune is spent on the purchase of a beautiful wall or floor covering material only to find that an insufficient amount of adhesive had been purchased to properly make the installation. The installer tried to "stretch" out the adhesive in order to complete the job. Most probably the number one cause of a job not holding up was that an inadequate amount of adhesive was used. For instance, it is possible that when applying the adhesive the correct size trowel was used as far as the notch sizes were concerned. However, you must remember that continuous use of a trowel on a concrete surface will wear the trowel down and change the notch size. Also, adhesive will tend to accumulate in the notches and dry, thereby partially clogging them and changing the notch size. Another factor is that no two people hold a trowel at the same angle. This will either increase or cut down on the notch size. Of course, if the incorrect trowel size was used from the beginning, it is obvious that you will encounter many difficulties. Always follow the adhesive manufacturer's recommendations.

By the same token, you may use too much adhesive, which would result in a failure of the installation. Usually this occurs when you try to fill too large a gap with adhesive. The strength of an adhesive is usually measured by the degree of adhesion to the substrate and the adherend, not to the cohesiveness within the adhesive itself. Therefore, weaker bonds will most generally occur when too much adhesive is used.

Adhesive Too Dry Before Assembly Made. There are several factors to consider which may cause poor bonds as a result of excessively dried adhesive films: the size of the bead from a cartridge; the size of the notches on the trowel; the nature of the substrates; ambient temperature and humidity conditions; and too long a waiting period before assembly.

1. **Trowel and Cartridge Application.** The open or working time of most adhesives is one of the variables that is generally controlled by the manufacturer during his formulation. He adjusts this to meet the requirements of the nature of the application. For example, a bead extruded from a cartridge will generally have a different open or working time than a troweled material. Thus, he adjusts his formulation to meet the particular type of installation. He will recommend to you on the label the exact bead size or notch trowel size to use for the optimum results. Therefore, you can see that if you use a smaller size of bead (Fig. 9-1) or notch than he has recommended, you can greatly reduce the open or working time, thereby causing the adhesive to be too dry to develop a satisfactory bond.

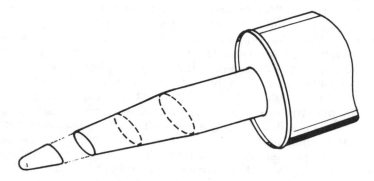

Fig.9-1: The size of the bead can be controlled by the way the tube nozzle is cut. The more that is cut from the tip of the nozzle, the larger the bead.

2. **Nature of Substrates.** Some surfaces have much different porosity or absorbency rates than others. For example, there are a lot of different types of particleboard made from different wood species. Some wood species are more porous than others and will allow the solvents or water, in the case of emulsion formulations, to be absorbed more readily, thereby shortening the open or working times. Steel troweled concrete is also far less porous or absorbent than regular masonry concrete and, as a result, will not permit solvents to dissipate as rapidly as conventional masonry surfaces. Metal or glass surfaces will not permit the drying or release of solvent as quickly as will wood or similar surfaces. Thus, you should consider this porosity or absorbency factor in your substrates and be guided accordingly. In general, the open or working time will not be nearly as long on the porous type of materials as it will be on the nonporous materials.

3. **Temperature and Humidity Conditions.** Generally speaking, higher temperatures will result in a more rapid release of the solvents in a given formulation, thereby shortening the open or working times. However, humidity also plays an important role. Even if the temperatures are high, under high humidity conditions, solvent release could be extremely slow, resulting in a much longer open or working time. Of course, the presence of air flow or wind would also have an effect on the open times, tending to shorten them considerably.

4. **Waiting Period Too Long.** While any one or a combination of the above factors could create problems for you, probably none are as common as just plain waiting too long after the adhesive is applied to make the assembly.

On any job, it is wise to occasionally pull apart an assembly shortly after it was made just to make sure that you have wet adhesive transfer from one surface to another. If not, you can be quite certain that the adhesive had dried too long before the assembly was made and that you are going to have problems.

Improperly Positioned, Clamped, or Braced During Drying. The consequences of any of these conditions are fairly obvious. Extra care during these operations will help assure you of avoiding problems in the future.

Positioning, of course, is very important. In the case of contact cements, you do not have a second chance to position the substrates, since bond is instantaneous and you cannot shift or adjust once you have brought the assembly together. In other types of installations, improper positioning can result in too much adjustment, thereby wiping adhesive from one of the substrates and providing you with a "starved" joint or bond with inadequate adhesive.

Improper clamping or bracing can also create problems. Not having the assembly in close enough contact during drying or curing will cause an unsatisfactory bond.

Bonded Assembly Improperly Set, Cured, or Dried. Most of us are guilty of being impatient from time to time, and there is something about gluing or bonding that brings out the worst in many of us. Many good bonds are destroyed by moving the assembly about before the adhesive or glue has had an opportunity to adequately cure, dry, or set. Unfortunately, there is no good rule of thumb on this situation, since there are so many different kinds of glues and adhesives, all of which have different characteristics relative to the rate of curing or drying. Some good wood glues, for example, will adequately cure in a matter of minutes, while a so-called fast-drying rubber mastic might take 12 to 24 hours to develop a satisfactory bond. Again, the best advice is to follow the directions on the label or, if possible, obtain a more detailed data or instruction sheet from your dealer or directly from the adhesive manufacturer.

Adhesive or Glue Did Not Meet Performance Requirements of Bonded Assembly

Where and how is the final bonded assembly going to be used? This is a key question and a topic that can create all kinds of problems for you if it is not seriously considered. All too often, the final assembly will be used outdoors in all kinds of weather, yet a waterproof adhesive was not used on it. Do not be mislead by terms such as water-resistant or waterproof on the label. Many manufacturers are inclined to take some poetic license in describing their products and they get carried away. Make sure that if it is going to be used outdoors, the label not only states "waterproof," but also describes the conditions under which it can be used and will perform. Most of the wood glues, for example, are not waterproof and should not be exposed to exterior conditions.

Temperature is another consideration. Many adhesives are exposed to temperatures, particularly in roof and ceiling applications, that they were never designed to tolerate. Some adhesives have poor metal adhesion yet bond well to wood and other surfaces. Strengths are also to be considered. It makes no sense to use, for example, an epoxy adhesive that will develop thousands of pounds per square inch strength when even one of the substrates will fracture within itself with only a couple of hundred pounds per square inch of shear force.

We suggest that you review the data in Chapters 2 and 7 regarding the initial selection of adhesives. This will help remind you to consider some of the conditions existing where the adhesive will be used. However, common sense and practicality will many times point you in the right direction and help you to avoid the selection of an adhesive or glue that will not perform under the conditions you require.

Have Not Read Manufacturer's Instructions

There is an old saying that states: *When everything else fails, read the directions* (Fig. 9-2). Many of us are guilty of complying with this bit of sarcasm from time to time. However, a thorough reading of the manufacturer's label and directions would go a long way toward preventing some of the trouble areas that are occasionally encountered. After all, the cost of the adhesive or glue that you are using is only a minute part of the total cost of your project in terms of time, labor, and materials. Therefore, the little time it takes you to familiarize yourself with the label copy and directions is just some added insurance for you to achieve a better and more professional job.

Fig. 9-2: One of the major causes of adhesive problems is the failure to read the container instructions. As mentioned in Chapter 1, manufacturers' labels carry a great deal of good information about their products.

Glossary

Abrasion resistance. Resistance to wear resulting from mechanical action on a surface.

Accelerated aging. A set of laboratory conditions designed to produce in a short time the results of normal aging. Usual factors included are temperature, light, oxygen, and water. In recent years, the adhesives industry has come to rely more and more on the "oxygen bomb" test as an indicator or relative life expectancy of a given formulation.

Accelerated weathering. A set of laboratory conditions to simulate in a short time the effects of natural weathering. Most adhesives are generally not subjected to the conditions that are normally considered under weathering tests.

Accelerator. An ingredient used in small amounts to speed up the action of a hardener.

Acetone. A very volatile solvent that is particularly useful for cleaning metal substrates.

Adhere. To cause two surfaces to be held together by adhesion.

Adherend. A body which is held to another body by an adhesive.

Adhesion. The state in which two surfaces are held together by interfacial forces which may consist of valence forces or interlocking action, or both.

Adhesion, mechanical. Adhesion between surfaces in which the adhesive holds the parts together by interlocking action.

Adhesion, specific. Adhesion between surfaces which are held together by valence forces of the same type as those which give rise to cohesion.

Adhesive. A substance capable of holding materials together by surface attachment. Same as **cement.**

Adhesive, assembly. An adhesive that can be used for bonding parts together such as in the manufacture of a boat, airplane, furniture, and the like.

Adhesive, cold-setting. An adhesive that sets at temperatures below 68°F (20°C).

Adhesive, contact. An adhesive that is apparently dry to the touch and which will adhere to itself instantaneously upon contact; also called contact bond adhesive or dry bond adhesive.

Adhesive, dispersion. A two phase system in which one phase is suspended in a liquid.

Adhesive failure. Type of failure characterized by pulling the adhesive loose from the adherend.

Adhesive, foamed. An adhesive, the apparent density of which has been decreased substantially by the presence of numerous gaseous cells dispersed throughout its mass. Same as **cellular adhesive.**

Adhesive, heat activated. A dry adhesive film that is rendered tacky or fluid by application of heat or heat and pressure to the assembly.

Adhesive, hot melt. An adhesive that is applied in a molten state and forms a bond on cooling to a solid state.

Adhesive, hot-setting. An adhesive that requires a temperature at or above 100°C (212°F) to set it.

Adhesive, intermediate temperature setting. An adhesive that sets in the temperature range of 31° to 99°C (87° to 211°F). Same as **intermediate temperature setting.**

Adhesive, multiple layer. A film adhesive with a different adhesive composition on each side; designed to bond dissimilar materials such as the core to face bond of a sandwich composite.

Adhesive, pressure-sensitive. A viscoelastic material which in solvent-free form remains permanently tacky. Such a material will adhere instantaneously to most solid surfaces with the application of very slight pressure.

Adhesive, room temperature setting. An adhesive that sets in the temperature range of 20° to 30°C (68° to 86°F).

Adhesive, separate application. A term used to describe an adhesive consisting of two parts, one part being applied to one adherend and the other part to the other adherend and the two brought together to form a joint.

Adhesive, solvent. An adhesive having a volatile organic liquid as a vehicle.
Adhesive, solvent activated. A dry adhesive film that is rendered tacky just prior to use by application of a solvent.
Adsorption. The action of a body in condensing and holding gases and other materials at its surface.
Aging. The progressive change in the chemical and physical properties of a sealant or adhesive.
Alligatoring. Cracking of a surface into segments so that it resembles the hide of an alligator.
Ambient temperature. Temperature of the air surrounding the object under construction.
Anaerobic. Adhesives that cure in the absence of oxygen.
A-stage. An early stage in the reaction of certain thermosetting resins in which the material is fusible and still soluble in certain liquids. Same as **resol.**
Assembly. A group of materials or parts, including adhesive, which has been placed together for bonding or which has been bonded together.
Asphalt. Naturally occurring mineral pitch or bitumen.

Batch. The manufactured unit or a blend of two or more units of the same formulation and processing.
B-stage. An intermediate stage in the reaction of certain thermosetting resins in which the material softens when heated and swells when in contact with certain liquids, but may not entirely fuse or dissolve. The resin in an uncured thermosetting adhesive is usually in this stage. Same as **resitol.**
Binder. A component of an adhesive composition that is primarily responsible for the adhesive forces which hold two bodies together.
Blister. An elevation of the surface of an adherend, somewhat resembling in shape a blister on the human skin; its boundaries may be indefinitely outlined and it may have burst and become flattened.
Blocking. An undesired adhesion between touching layers of a material such as occurs under moderate pressure during storage or use.
Bond. The attachment at an interface between substrate and adhesive, or sealant.
Bond. To join materials together using an adhesive.
Bond face. The part or surface of a building component which serves as a substrate for an adhesive.
Bond strength. The unit load applied in tension, compression, flexure, peel, impact, cleavage, or shear, that is required to break an adhesive assembly with failure occurring in or near the plane of the bond.

C-stage. The final stage in the reaction of certain thermosetting resins in which the material is relatively insoluble and infusible. Certain thermosetting resins in a fully cured adhesive layer are in this stage. Same as **resite.**
Catalyst. Substance added in small quantities to promote a reaction, while remaining unchanged itself.
Cellular material. A material containing many small cells dispersed throughout it. The cells may be either open or closed.
Checking. The formation of slight breaks or cracks in the surface of an adhesive.
Chemical cure. Curing by chemical reaction. Usually involves the cross-linking of a polymer.
Closed cell. A cell enclosed by its walls and therefore not connected to other cells.
Coefficient of expansion. The coefficient of linear expansion is the ratio of the change in length per degree to the length at $0°C$.
Cohesion. The molecular attraction which holds the body of an adhesive together. The internal strength of an adhesive.
Cohesive failure. The failure characterized by pulling the body of the adhesive apart.
Cohesive strength. The ability of the adhesive to stick within itself during the wet stage.
Cold pressing. A bonding operation in which an assembly is subjected to pressure without the application of heat.
Condensation. A chemical reaction in which two or more molecules combine with the separation of water or some other simple substance. If a polymer is formed, the process is called **polycondensation.**
Consistency. That property of a liquid adhesive by virtue of which it tends to resist deformation.
Crazing. Fine cracks that may extend in a network on or under the surface of or through a layer of adhesive.
Creep. The deformation of a body with time under constant load. Also called **cold flow.**
Cure. To set up or harden by means of a chemical reaction.
Cure time. Time required to effect a complete cure at a given temperature.

Curing agent. A chemical which is added to effect a cure in a polymer. Same as **hardener.**

Delamination. The separation of layers in a laminate because of failure of the adhesive, either in the adhesive itself or at the interface between the adhesive and the adherend, or because of cohesive failure of the adherend.

Diluent. An ingredient usually added to an adhesive to reduce the concentration of bonding materials.

Doctor (bar or blade). Device that controls the amount of adhesive applied.

Dry. To change the physical state of an adhesive on an adherend by the loss of solvent constituents by evaporation or absorption, or both.

Elasticity. The ability of a material to return to its original shape after removal of a load.

Elastomer. A rubbery material which returns to approximately its original dimensions in a short time after a relatively large amount of deformation.

Emulsion. A dispersion of fine particles in water.

Exothermic. A chemical reaction which gives off heat.

Extender. An organic material used to increase the volume and lower the cost of an adhesive.

Failure, adhesive. Rupture of an adhesive bond such that the separation appears to be at the adhesive adherend interface.

Fatigue failure. Failure of a material due to rapid cyclic deformation.

Filler. Finely ground material added to an adhesive to change or improve certain properties.

Fillet. That portion of an adhesive which fills the corner or angle formed where two adherends are joined.

Flow. Movement of an adhesive during the bonding process before the adhesive is set.

Gel. A semisolid system consisting of a network of solid aggregates in which liquid is held.

Glue. Originally, a hard gelatin obtained from hides, tendons, cartilage, bones, etc., of animals. Also, an adhesive prepared from this substance by heating with water. Through general use the term is now synonymous with the term "adhesive."

Glue line. The layer of adhesive which attaches two adherends. Same as **bond line.**

Green strength. This refers to the relative cohesive strength an adhesive, glue, or mastic has in the wet state. Same as green grab or initial tack. See also **tack.**

Gum. Any of a class of colloidal substances, exuded by or prepared from plants, sticky when moist, composed of complex carbohydrates and organic acids, which are soluble or swell in water.

Hardener. A substance or mixture of substances added to an adhesive to promote or control the curing reaction by taking part in it. The term is also used to designate a substance added to control the degree of hardness of the cured fill. Same as a **curing agent.** See also **catalyst.**

Inhibitor. A substance that slows down chemical reaction. Inhibitors are sometimes used in certain types of adhesives to prolong storage or working life. Same as a **retarder.**

Interface. The common boundary surface between two substances.

Joint. The location at which two adherends are held together with a layer of adhesive.

Joint, lap. A joint made by placing one adherend partly over another and bonding together the overlapped portions.

Joint, scarf. A joint made by cutting away similar angular segments of two adherends and bonding the adherends with the cut areas fitted together.

Joint, starved. A joint that has an insufficient amount of adhesive to produce a satisfactory bond.

"J" Roller. A hand roller used to apply pressure on a bonded surface such as a plastic laminate.

Laminate, (noun). A product made by bonding together two or more layers of material or materials.

Laminate, (verb). To unite layers of material with adhesive.

Laminated, cross. A laminate in which some of the layers of material are oriented at right angles to the remaining layers with respect to the grain or strongest direction in tension.

Laminated, parallel. A laminate in which all the layers of material are oriented approximately parallel with respect to the grain or strongest direction in tension.

Legging. The drawing of filaments or strings when adhesive-bonded substrates are separated.

Matrix. The part of an adhesive which surrounds or engulfs embedded filler or reinforcing particles and filaments.

Modifier. Any chemically inert ingredient added to an adhesive formulation that changes its properties.

Monomer. A relatively simple compound which can react to form a polymer.

Mucilage. An adhesive prepared from a gum and water. Also in a more general sense, a liquid adhesive which has a low order of bonding strength.

Open time. Time interval between when an adhesive is applied and when it becomes no longer workable. Same as **working time.** Same as **open assembly time.**

Oxygen bomb test. A special aging test given to adhesives. Five hundred hours exposure to the condition in this test generally indicates whether a product will provide a good deal of service over a long range period of time.

Paste. An adhesive composition having a characteristic plastic-type consistency, that is, a high order or yield value, such as that of a paste prepared by heating a mixture of starch and water and subsequently cooling the hydrolyzed product.

Peel test. A test of an adhesive using one rigid and one flexible substrate. The flexible material is folded back (usually 180°) and the substrates are peeled apart. Strength is measured in pounds per inch of width.

Penetration. The entering of an adhesive into an adherend.

Permanence. The resistance of an adhesive bond to deteriorating influences.

Permanent set. The amount of deformation which remains in an adhesive after removal of a load.

Phenolic resin. A thermosetting resin. Usually formed by the reaction of a phenol with formaldehyde.

Pick-up roll. A spreading device where the roll for picking up the adhesive runs in a reservoir of adhesive.

Pitch. The residue which remains after the distillation of oil and so forth from raw petroleum.

Plasticity. A property of adhesives that allows the material to be deformed continuously and permanently without rupture upon the application of a force that exceeds the yield value of the material.

Plasticizer. A material incorporated in an adhesive to increase its flexibility, workability, or distensibility. The addition of the plasticizer may cause a reduction in melt viscosity, lower the temperature of the second-order transition, or lower the elastic modulus of the solidified adhesive.

Polymer. A compound formed by the reaction of simple molecules having functional groups which permit their combination to proceed to high molecular weights under suitable conditions. Polymers may be formed by polymerization (addition polymer) or polycondensation (condensation polymer). When two or more monomers are involved, the product is called a *copolymer.*

Polymerization. A chemical reaction in which the molecules of a monomer are linked together to form large molecules whose molecular weight is a multiple of that of the original substance. When two or more monomers are involved, the process is called *copolymerization* or *heteropolymerization.*

Post cure, (noun). A treatment (normally involving heat) applied to an adhesive assembly following the initial cure to modify specific properties.

Post cure, (verb). To expose an adhesive assembly to an additional cure, following the initial cure, for the purpose of modifying specific properties.

Pot life. Same as **working life.**

Primer. A coating applied to a surface, prior to the application of an adhesive, to improve the performance of the bond.

Release paper. A sheet, serving as a protectant and/or carrier for an adhesive film or mass, which is easily removed from the film or mass prior to use.

Resin. A solid, semisolid, or pseudosolid organic material that has an indefinite and often high molecular weight, exhibits a tendency to flow when subjected to stress, usually has a softening or melting range, and usually fractures conchoidally.

Rosin. A resin obtained as a residue in the distillation of crude turpentine from the sap of the pine tree (gum rosin) or from an extract of the stumps and other parts of the tree (wood rosin).

Self-vulcanizing. Pertaining to an adhesive that undergoes vulcanization without the application of heat. Same as **self-curing.**

Set. To convert an adhesive into a fixed or hardened state by chemical or physical action, such as condensation, polymerization, oxidation, vulcanization, gelation, hydration, or evaporation of volatile constituents.

Shrinkage. Percentage weight loss under specified conditions.

Sizing. The process of applying a material on a surface in order to fill pores and thus reduce the absorption of the subsequently applied adhesive or coating or to otherwise modify the surface properties of the substrate to improve the adhesion. Also, the material used for this purpose. The latter is called a *size.*

Slippage. The movement of adherends with respect to each other during the bonding process.

Solids content. The percentage by weight of the nonvolatile matter in an adhesive.

Solvent. Liquid in which another substance can be dissolved.

Spread. The quantity of adhesive per unit joint area applied to an adherend, usually expressed in points of adhesive per thousand square feet of joint area. *Single spread* refers to application of adhesive to only one adherend of a joint. *Double spread* refers to application of adhesive to both adherends of a joint.

Squeeze out. Adhesive pressed out at the bond line due to pressure applied on the adherends.

Storage life. The period of time during which a packaged adhesive can be stored under specified temperature conditions and remain suitable for use. Sometimes called *shelf life.*

Strength, dry. The strength of an adhesive joint determined immediately after drying under specified conditions or after a period of conditioning in the standard laboratory atmosphere.

Strength, wet. The strength of an adhesive joint determined immediately after removal from a liquid in which it has been immersed under specified conditions of time, temperature, and pressure.

Stress. Force per unit area, usually expressed in pounds per square inch (psi).

Stress relaxation. Reduction in stress in a material which is held at a constant deformation for an extended time.

Stringiness. The property of an adhesive that results in the formation of filaments or threads when adhesive transfer surfaces are separated.

Structural adhesive. A bonding agent used for transferring required loads between adherends exposed to service environments typical for the structure involved.

Substrate. A material upon the surface of which an adhesive-containing substance is spread for any purpose, such as bonding or coating. A broader term than adherend.

Surface preparation. A physical and/or chemical preparation of an adherend to render it suitable for adhesive joining. Same as **adherend preparation** or **prebond preparation.**

Tack. The property of an adhesive that enables it to form a bond of measurable strength immediately after adhesive and adherend are brought into contact under low pressure.

Tack, dry. The property of certain adhesives, particularly nonvulcanizing rubber adhesives, to adhere on contact to themselves at a stage in the evaporation of volatile constituents, even though they seem dry to the touch. Same as **aggressive tack.**

Tack range. The period of time in which an adhesive will remain in the tacky-dry condition after application to an adherend, under specified conditions of temperature and humidity.

Tackiness. The stickiness of the surface of a sealant or adhesive.

Tack-dry. Pertaining to the condition of an adhesive when the volatile constituents have evaporated or been absorbed sufficiently to leave it in a desired tacky state.

Tear strength. The load required to tear apart a sealant specimen.

Teeth. The resultant surface irregularities or projections formed by the breaking of filaments or strings which may form when adhesive-bonded substrates are separated.

Telegraphing. A condition in a laminate or other type of composite construction in which irregularities, imperfections, or patterns of an inner layer are visibly transmitted to the surface.

Temperature, curing. The temperature to which an adhesive or an assembly is subjected to cure the adhesive.

Temperature, drying. The temperature to which an adhesive on an adherend or in an assembly, or the assembly itself is subjected to dry the adhesive.

Temperature, maturing. The temperature, as a function of time and bonding condition, that produces desired characteristics in bonded components.

Temperature, setting. The temperature to which an adhesive or an assembly is subjected to set the adhesive.

Tensile strength. Resistance of a material to a tensile force (a stretch). The cohesive strength of a material expressed in psi.

Thermoplastic, (adjective). Capable of being repeatedly softened by heat and hardened by cooling.

Thermoplastic, (noun). A material that will repeatedly soften when heated and harden when cooled.

Thermoset. Pertaining to the state of a resin in which it is relatively infusible.

Thermoset. A material that will undergo or has undergone a chemical reaction by the action of heat, catalysts, ultraviolet light, etc., leading to a relatively infusible state.

Thermosetting. Having the property of undergoing a chemical reaction by the action of heat, catalysts, ultraviolet light, etc., leading to a relatively infusible state.

Thinner. A volatile liquid added to an adhesive to modify the consistency or other properties.

Thixotropic. Nonsagging. A material which maintains its shape unless agitated. A thixotropic sealant can be placed in a joint in a vertical wall and will maintain its shape without sagging during the curing process.

Time, assembly. The time interval between the spreading of the adhesive on the adherend and the application of pressure or heat, or both, to the assembly. Same as **closed assembly time.**

Time, curing. The period of time during which an assembly is subject to heat or pressure, or both, to cure the adhesive.

Time, drying. The period of time during which an adhesive on an adherend or an assembly is allowed to dry with or without the application of heat or pressure, or both.

Time, joint conditioning. The time interval between the removal of the joint from the conditions of heat or pressure, or both, used to accomplish bonding and the attainment of approximately maximum bond strength. Sometimes called **joint aging time.**

Time, setting. The period of time during which an assembly is subjected to heat or pressure, or both, to set the adhesive.

Toxic. Poisonous or dangerous to humans by swallowing, inhalation, or contact resulting in eye or skin irritation.

Ultimate elongation. Elongation at failure.

Ultraviolet light. Part of the light spectrum. Ultraviolet rays can cause chemical changes in rubbery materials.

Urethane. A family of polymers ranging from rubbery to brittle. Usually formed by the reaction of a di-isocyanate with a hydroxyl.

Vehicle. The liquid component of a material.

Viscosity. The ratio of the shear stress existing between laminae of moving fluid and the rate of shear between these laminae.

Vulcanization. A chemical reaction in which the physical properties of a rubber are changed in the direction of decreased plastic flow, less surface tackiness, and increased tensile strength by reacting it with sulfur or other suitable agents.

Vulcanize. To subject to vulcanization.

Warp. A significant variation from the original, true, or plane surface.

Webbing. Filaments or threads that may form when adhesive transfer surfaces are separated.

Wood, built-up laminated. An assembly made by joining layers of lumber with mechanical fastenings so that the grain of all laminations is essentially parallel.

Wood, glued laminated. An assembly made by bonding layers of veneer or lumber with an adhesive so that the grain of all laminations is essentially parallel.

Wood failure. The rupturing of wood fibers in strength tests on bonded specimens, usually expressed as the percentage of the total area involved which shows such failure.

Working life. The period of time during which an adhesive, after mixing with catalyst, solvent, or other compounding ingredients, remains suitable for use. Same as **pot life.**

Yield value. The stress (either normal or shear) at which a marked increase in deformation occurs without an increase in load.

Index